高等职业教育大数据技术专业系列教材

JavaWeb 数据可视化开发教程

主　编　张巍然
副主编　陈晓慧　郭 琳　肖 梅
　　　　侯艳红
主　审　魏晓艳　郭立文

西安电子科技大学出版社

内 容 简 介

本书采用工作手册式形式，以数据可视化的实际工作岗位需求为目标，将全书划分为 7 章：前 6 章的每章都对应一个核心内容、多个案例或实战演练，确保读者能够系统掌握 ECharts 库、SSM 框架的知识和技能；第 7 章是一个综合项目，帮助学生将所学知识转化为实际能力，提升职业竞争力。

本书通俗易懂，突出实用性，具有系统性的知识结构、针对性的入门实验以及大量的项目案例；采用目标导向的教学方法，通过教学设计引导读者自主探索和深入学习；根据内容特点融入思政教育，培养学生的社会责任感和职业道德，为未来的职业发展奠定坚实基础。

本书可作为高等学校计算机大类相关专业的教材，也可作为数据可视化、大数据应用、Java EE 开发等相关岗位技术人员的学习参考书。

图书在版编目（CIP）数据

JavaWeb 数据可视化开发教程 / 张巍然主编. -- 西安 : 西安电子科技大学出版社, 2025. 1. -- ISBN 978-7-5606-7494-0

Ⅰ. TP312.8

中国国家版本馆 CIP 数据核字第 2025XG3728 号

策　　划　高　樱
责任编辑　高　樱
出版发行　西安电子科技大学出版社（西安市太白南路 2 号）
电　　话　（029）88202421　88201467　　邮　　编　710071
网　　址　www.xduph.com　　电子邮箱　xdupfxb001@163.com
经　　销　新华书店
印刷单位　咸阳华盛印务有限责任公司
版　　次　2025 年 1 月第 1 版　　2025 年 1 月第 1 次印刷
开　　本　787 毫米×1092 毫米　1/16　　印 张 20
字　　数　475 千字
定　　价　54.00 元

ISBN 978-7-5606-7494-0

XDUP 7795001-1

***　如有印装问题可调换　***

前　言

在云计算和大数据时代的背景下，人才是数字经济发展的关键因素，需要加快建设结构多元、层次合理的人才队伍，培养高技能人才并发挥人才优势。本书以贯彻党的教育方针，落实立德树人为根本任务，以满足大数据分析、Java EE 开发等岗位需求为目标，紧密结合企业实际介绍了数据可视化的基本原理、方法和实践技巧，帮助读者掌握理论知识，提高实际操作能力。

本书共 7 章，详细讲解了基于 JavaWeb 的数据可视化技术。第 1 章概述了 JavaWeb 数据可视化技术；第 2 章详细介绍了 ECharts 基础图表，包括快速上手操作案例以及柱状图、折线图、饼图、散点图等基本图表的讲解，配合案例和实战演练介绍 ECharts 的综合应用；第 3 章进一步讲解了 ECharts 在仪表盘和漏斗图等图表中的应用，以及标题、图例、网格、坐标轴、工具箱、提示框、标记点和标记线等常用组件的使用方法；第 4 章通过知识讲解和具体案例介绍了 Spring 框架，包括 Spring Bean 的配置方法、依赖注入和 Spring 面向切面编程(AOP)等；第 5 章详细讲解了 Spring MVC 的相关知识，涵盖了快速上手案例、MVC 核心组件、Controller 的编写、请求接收数据、JSON 数据的转换处理以及 RESTful API 的支持等核心技术；第 6 章介绍了 MyBatis 的使用方法，涉及快速上手案例、配置文件、API、映射、动态 SQL、整合开发等内容；第 7 章介绍了一个智慧农业数据分析平台的项目实例，通过各功能模块的实现，让读者能够将之前章节学到的理论知识和技能综合应用到一个具体的项目中。

编写本书时，数据可视化课题组经过多次调研和讨论，明确教材定位和目标读者，制订了详细的大纲和目录。本书在内容设计上遵循技术技能人才成长规律，首先明确知识目标、技能目标和思政目标，确保学生能够清晰地了解重点；接着对理论知识概括叙述，并通过快速上手案例帮助学生入门体验和了解核心技术；再详细介绍相关理论知识并结合大量的案例或专门的实战演练，确保学生能够系统掌握知识技能；最后安排了强化练习、进一步学习建议和考核评价 3 个模块，帮助学生巩固和加深对知识点的理解，深入拓展知识和技能，全面评估学习成效，为未来的职业发展奠定坚实基础。

本书通俗易懂，突出实用性，整体上遵循从理论到实践、从基础到高级的循序渐进的学习路径；采用目标导向的教学方法，引导学生有的放矢地学习，增强学习效果；教材注重实践性，通过快速上手、案例分析以及实战演练相结合的方式，使学生能够迅速掌握并应用所学知识，在实际操作中加深理解，提高应用能力；提供进一步学习建议，鼓励学生自主探索和深入学习，将思政目标融入专业学习，培养学生的社会责任感和职业道德，帮助学生实现知识、技能和素养的全方位发展。

本书由陕西国防工业职业技术学院张巍然担任主编，陈晓慧、郭琳、肖梅、侯艳红担任副主编，其中郭琳撰写第 1、2 章，陈晓慧撰写第 3 章，张巍然撰写第 4、5 章和第 6.4～6.7 节，侯艳红撰写第 6.1～6.3 节，肖梅撰写第 7 章，侯艳红负责编写本书的习题并参与编写第 6 章，全书由张巍然、陈晓慧统稿，由魏晓艳、郭立文担任本书的主审并负责其他工作。在编写本书的过程中，我们参考了许多文献，在此对相关作者表示衷心的感谢；对所有帮助和支持本书出版的领导、同事和出版社老师表示由衷的感谢。

本书提供有配套的电子课件、教学大纲、电子教案、案例源码等资源，需要者可登录西安电子科技大学出版社官网下载。

由于编者水平有限，书中难免有不足之处，恳请各位专家和读者批评指正。

编　者

2024 年 8 月于西安

目　录

第 1 章 概 述

学习目标

目标类型	目 标 描 述
知识目标	• 理解 Java Web 技术的基本概念和架构 • 了解数据可视化的基本流程和关键环节 • 掌握基于 Java Web 的数据可视化实现过程
技能目标	• 培养学生使用搜索引擎、技术论坛、官方文档等资源来学习和解决问题的能力 • 掌握在 ECharts 官方网站下载相应版本的方法 • 掌握在 webpack 项目中通过 npm 命令安装 ECharts 的方法
思政目标	• 不断学习新知识、新技术，养成终身学习的习惯，为个人的职业发展奠定坚实的基础 • 了解国内外相关工具技术的应用和发展趋势，提升国际视野，培养国际竞争意识 • 理解数据可视化在社会管理、公共决策、教育科研等领域的应用，利用所学知识为社会发展作出贡献，提升社会参与意识

知识技能储备

1.1 JavaWeb 数据可视化

随着大数据和分析技术的不断进步，人们期望快速收集、分析和理解数据的实用价值，并将数据转换为图形，如折线图、柱状图、饼图、文字云和地图等，以通过数据可视化的形式来揭示数据中的变化趋势和异常，从而提高人们的洞察力和决策质量。数据可视化的具体作用如下：

(1) 通过图形找到数据的某些特征，分析其中所隐含的规律。图 1-1 展示了某地区一年内每月平均温度数据，由此可以分析得到该地区的气候变化规律，即呈现出强烈的季节性趋势：冬季平均温度约为 6.3℃，而夏季平均温度约为 23.3℃；

春季到夏季，温度以每月约 3.3℃的速度稳步上升，而秋季则以每月约 3℃的速度逐渐下降；夏季通常会出现高温峰值，也会因各种气候因素而产生波动。这些趋势和规律反映了该地区的气候特点，为预测未来气候变化提供了重要依据。

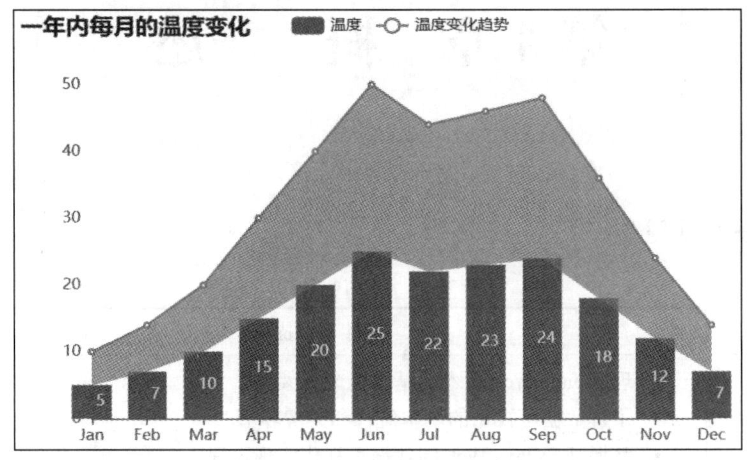

图 1-1 温度变化趋势

(2) 在数据分析和统计研究过程中，可以通过数据可视化找到不同要素之间存在的相互关系。图 1-2 为研究生平均每周学习时长与他们最终学业成绩的关系，可以看出一般情况下学习时间越长，考试成绩也越高。通常当其中一个变量增加而另一个变量也倾向于增加时，称这两个变量的关联性为正相关性；与此相反，当一个变量的增加通常伴随着另一个变量的减少时，称这两个变量之间的关联性为负相关性，例如随着工资的增加，失业率可能下降。但是关联性的因素往往不是唯一决定因素，例如人们的身高和体重之间通常存在正相关性，但这并不意味着身高是体重的唯一决定因素，因为遗传、饮食和生活方式等因素也对体重有影响。

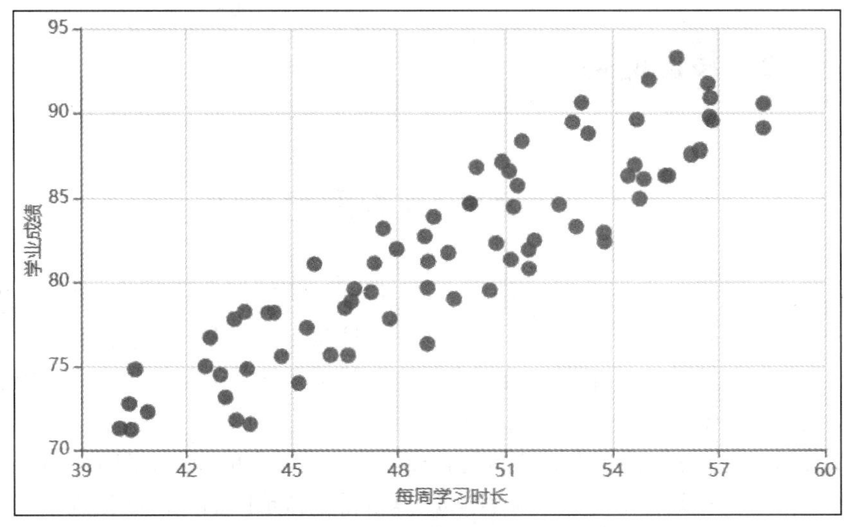

图 1-2 学业成绩与每周学习时长的关系

(3) 数据可视化不仅有助于发现问题，还提供了及时应对和优化决策的重要依据。当数据以图表形式呈现时，异常值、异常模式、不一致性和错误更容易被识别，这对于及时发现问题和采取行动至关重要。例如，在使用折线图分析网站流量数据时，很容易发现某个时间段的流量突然下降，这种异常模式表明了服务器故障、网站更新、网络问题等可能情况。通过快速识别这些异常情况，可以及时采取措施，如修复服务器、调整更新策略或启动危机管理等，以保持网站稳定运行和良好的用户体验。

传统的表格无法表达大量、复杂的数据信息且缺乏交互性，而且手动处理数据通常效率低下，尤其是人工逐一处理时，容易出错，通常限于简单的统计和图表制作，无法动态展示数据变化。JavaWeb 数据可视化的目标是为用户提供一种直观且交互的方式来展示和分析数据，通过使用如 ECharts 这样的工具提供丰富的图表功能，使数据更加直观易懂。JavaWeb 支持前后端分离的开发模式，后端负责处理业务逻辑，并能轻松连接到各种数据库，为数据可视化提供了丰富的数据源，而前端负责展示数据和用户交互，结合 Ajax 技术可以实现前后端的异步数据交互，从而实现动态数据的实时展示。

基于 JavaWeb 的数据可视化流程包括数据收集、数据预处理、数据存储、数据可视化设计、用户交互处理、部署与测试以及维护与优化等步骤。

(1) 数据收集：确定数据的可视化来源，这些数据可以来自数据库、文件、网络或其他数据服务。使用 Java 技术(如 JDBC、JPA、MyBatis 等)连接到数据源，并收集所需的数据。

(2) 数据预处理：对数据清洗、转换、聚合，以确保数据质量和一致性。

(3) 数据存储：数据存储是数据可视化流程中的一个关键步骤，它涉及将预处理后的数据保存到数据库或其他数据存储系统中，以便于后续的检索和分析。

(4) 数据可视化设计：确定需要展示的数据指标和图表类型，如柱状图、折线图、饼图等；使用前端技术(如 JSP、JSF、HTML、CSS、JavaScript 等)来创建用户界面；还需要选择合适的数据可视化库或框架，如 D3.js、ECharts、Highcharts、Google Charts 等，来辅助实现数据可视化到界面上。

(5) 用户交互处理：允许用户与可视化界面交互，如缩放、滚动、筛选等。捕捉用户的交互行为，并使用 Java 后端技术(如 Servlet、Spring MVC 等)处理用户的请求，并将数据传递给前端界面，根据用户的需求展示数据，动态更新可视化内容。

(6) 部署与测试：将开发完成的数据可视化应用部署到 Java Web 服务器上，如 Tomcat、JBoss、WebSphere 等。进行系统测试，确保功能正常且性能满足需求。

(7) 维护与优化：对数据可视化应用进行定期的维护和更新，确保数据的准确性和应用的稳定性；根据用户反馈和业务需求，对数据可视化流程进行优化和调整。

Java Web 技术的强大之处在于它能够提供跨平台的解决方案，并且能够与多种数据源和可视化库无缝集成。本书将使用基于 SSM 框架的 Java Web 技术结合 ECharts 库来实现数据可视化。

1.2 相关技术

1.2.1 数据可视化库 ECharts

ECharts(Enterprise Charts)是一个使用 JavaScript 实现的开源可视化库，由百度团队开发，主要用于在 Web 页面上展示数据。它可以流畅地运行在 PC 和移动设备上，兼容当前绝大部分浏览器，底层依赖轻量级的 Canvas 类库 ZRender，提供直观、生动、可交互、可高度个性化定制的数据可视化图表。ECharts 除了提供常规的折线图、柱状图、散点图、饼图、K 线图之外，还提供用于统计的盒形图，用于地理数据可视化的地图、热力图、线图，以及用于 BI(Business Intelligence，商业智能)的漏斗图和仪表盘，并且支持图与图之间的混搭。

获取 ECharts 的方式主要有以下 3 种。

方式一：直接下载。

在官网的下载页面 https://archive.apache.org/dist/echarts/中选择要下载的 ECharts 源代码；或者从 GitHub 仓库 https://github.com/apache/echarts 下载编译产物。下载后解压得到 ECharts，包括 echarts.min.js(压缩后的生产环境版本)和 echarts.js(未压缩的开发环境版本)。

方式二：npm 安装。

如果使用 npm，可以在命令行中运行以下命令来安装 ECharts：

```
npm install echarts --save
```

方式三：选择需要的模块，在线定制下载。

这种方式可自由选择所需图表、坐标系、组件进行打包下载，打包的源文件来自 npm mirror 镜像站，该源文件并不是 Apache 官方源代码和编译产物。具体方法如下：

(1) 在浏览器中输入官网地址 https://echarts.apache.org/zh/index.html，进入官网页面。

(2) 在官网页面，单击下载菜单中的“下载”选项，如图 1-3 所示。

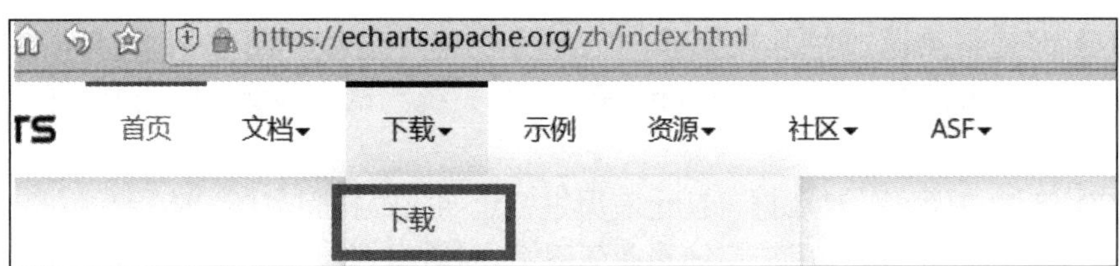

图 1-3 下载选项

(3) 出现如图 1-4 所示的下载页面后，单击“在线定制”按钮。

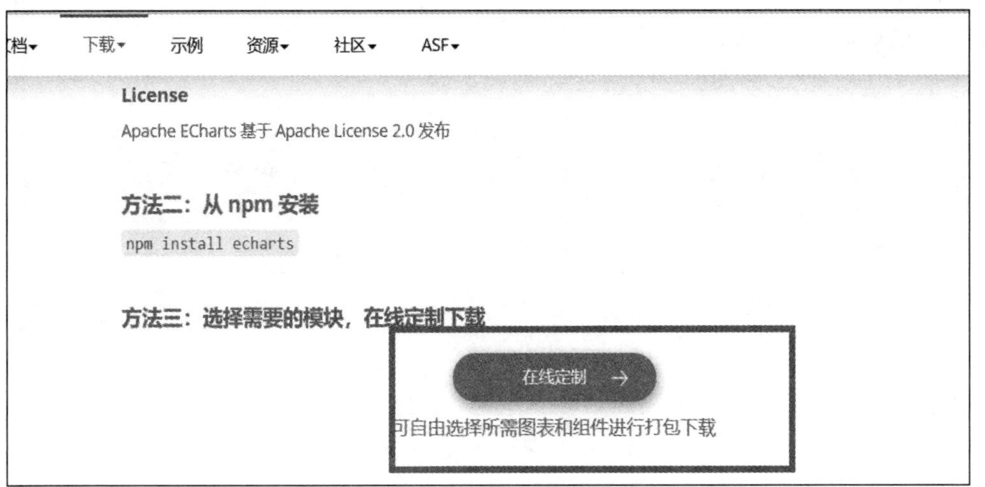

图 1-4　下载页面

(4) 进入如图 1-5 所示的在线定制页面，选择版本并勾选所需要的图表类型。

图 1-5　在线定制页面

(5) 在如图 1-6 所示的在线定制页面下方的坐标系和组件选择部分，勾选需要的坐标系和组件类型。

(6) 在如图 1-7 所示的其他选项中取消代码压缩，然后单击“下载”按钮，即可得到 echarts.js。

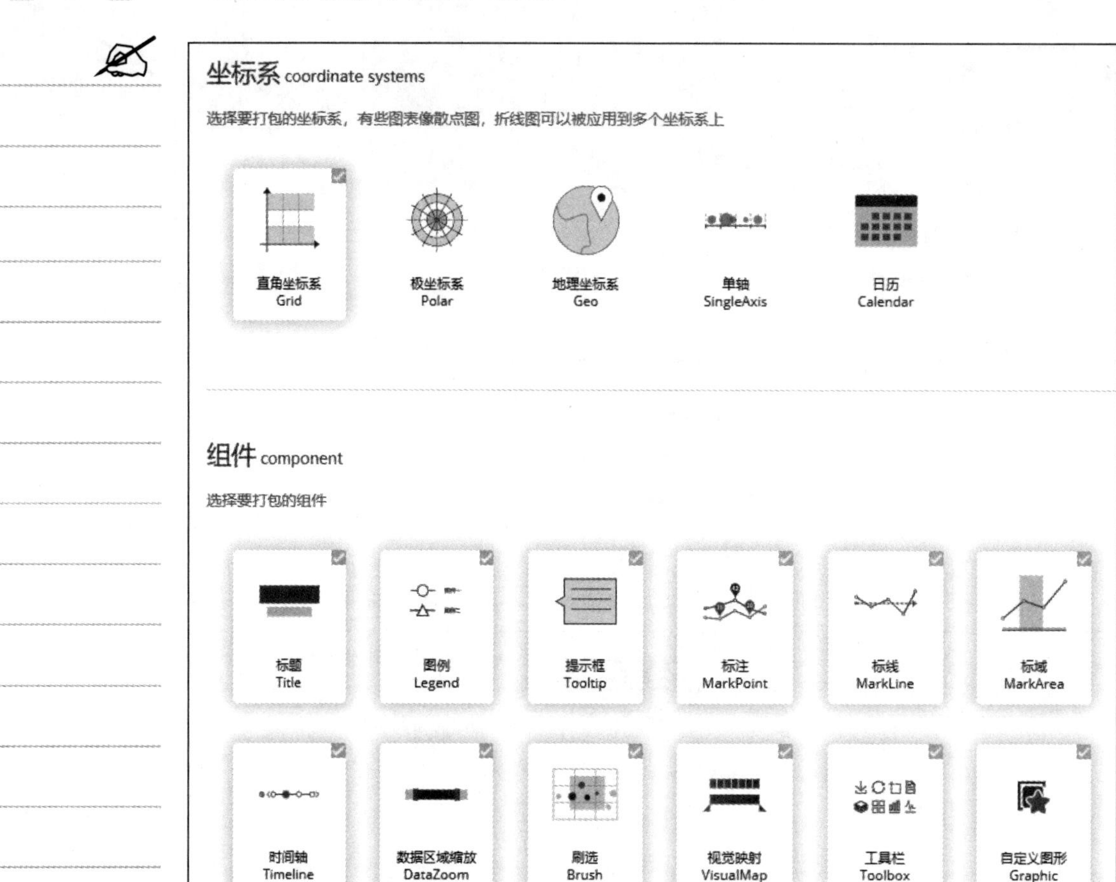

图 1-6　选择坐标系和组件类型

其它选项 others

☑ SVG 渲染

是否包括 SVG 渲染器，从而能支持使用 SVG 来绘制图表

☑ 工具集

是否在 echarts 对象上挂载常用工具集。一般都会挂载，除非对生成的文件的体积有苛求，并且不需要用这些工具集。

☐ 代码压缩

是否使用 UglifyJS 压缩后的代码，开发环境建议不压缩代码，代码压缩会去掉大部分常见的警告和错误提示。

下载

图 1-7　其他选项

1.2.2 SSM 框架技术

SSM 是 Spring、Spring MVC 和 MyBatis 这 3 种框架的缩写，这些框架通常一起使用，以创建能够处理复杂业务需求、组织良好且易于维护的应用程序。

Spring 是一个开源的 Java 企业级应用程序开发框架，旨在简化 Java 企业级应用的开发和维护工作，它最初由 Rod Johnson 设计并首次发布于 2003 年。Spring 框架的核心是它的 IoC(Inversion of Control，控制反转)和 AOP(Aspect Oriented Programming，面向切面编程)思想；IoC 允许开发者通过外部容器来控制对象的创建和依赖关系，而 AOP 则允许开发者将横切关注点(如日志、事务管理等)与业务逻辑分离。

Spring MVC 是 Spring 框架的一个模块，基于 Java 的 Web-MVC(Web-Model View Controller)设计模式，用于构建 Web 应用程序。它提供了用于开发 Web 应用程序的全套机制，包括请求处理、视图渲染、数据验证等；还提供了一系列特性，如数据绑定、表单验证、国际化等，以支持复杂的 Web 应用程序的开发。

MyBatis 是一个优秀的持久层框架，它支持普通的 SQL 语句，同时也支持复杂的存储过程以及高级映射。MyBatis 避免了几乎所有的 JDBC 代码、手动设置参数和获取结果集。其主要优点为：通过 XML 文件或注解的方式配置 SQL 语句，使得 SQL 逻辑与 Java 代码分离，这样可以使代码更加简洁并便于维护；MyBatis 支持动态 SQL，允许编写可重用的 SQL 语句；MyBatis 与 Spring 框架的整合可以使 JavaWeb 应用程序的开发变得更加高效和便捷。

强化练习

1. 简述 JavaWeb 数据可视化的实现步骤。
2. 简要介绍 ECharts 的作用和获取方法。
3. 解释 SSM 框架的作用以及各组成部分。

习题答案

进一步学习建议

可以进一步学习以下内容，以提高技能和加强理解：

(1) 学习基本图表类型，如常见的柱状图、折线图、饼图等。

(2) 建议深入学习基本图表的原理、特点和适用场景，以便在实际项目中灵活运用。

考核评价

<table>
<tr><td colspan="5">考核评价表</td></tr>
<tr><td colspan="2">姓名</td><td></td><td>班　级</td><td></td></tr>
<tr><td colspan="2">学号</td><td></td><td>考评时间</td><td></td></tr>
<tr><td colspan="2">评价主题及总分</td><td colspan="2">评价内容及分数</td><td>评分</td></tr>
<tr><td rowspan="3">1</td><td rowspan="3">知识考核
(30)</td><td colspan="2">能够初步描述 Java Web 技术的基本概念(10 分)</td><td></td></tr>
<tr><td colspan="2">举例描述数据可视化的基本流程(10 分)</td><td></td></tr>
<tr><td colspan="2">介绍 ECharts 的功能和作用(10 分)</td><td></td></tr>
<tr><td rowspan="3">2</td><td rowspan="3">技能考核
(40)</td><td colspan="2">正确下载和使用 ECharts 的不同版本，并说明源代码版本的优势(10 分)</td><td></td></tr>
<tr><td colspan="2">在 GitHub 上找到 ECharts 的 Release 版本，并能够在解压后的目录中找到 ECharts 库(20 分)</td><td></td></tr>
<tr><td colspan="2">展示使用搜索引擎、技术论坛、官方文档等资源来解决问题和学习的技巧(10 分)</td><td></td></tr>
<tr><td rowspan="3">3</td><td rowspan="3">思政考核
(30)</td><td colspan="2">通过笔记、博客或与他人交流的方式记录和分享学习成果(10 分)</td><td></td></tr>
<tr><td colspan="2">介绍数据可视化技术的发展趋势和应用场景(10 分)</td><td></td></tr>
<tr><td colspan="2">举例数据可视化技术在社会管理、公共决策、教育科研等领域的应用(10 分)</td><td></td></tr>
<tr><td colspan="4">评语：</td><td>汇总：</td></tr>
</table>

第 2 章　ECharts 基础图表

学习目标

目标类型	目 标 描 述
知识目标	• 掌握 ECharts 常用图表(柱状图、折线图、饼图、散点图、气泡图)的原理、特点和适用场景 • 熟悉 ECharts 的配置项和参数设置，能够根据需求进行合理的配置 • 理解 ECharts 的交互功能(如数据区域缩放、数据点 hover 等)及其在数据可视化中的作用
技能目标	• 能够根据实际需求选择合适的图表类型，并利用 ECharts 实现数据的可视化 • 掌握 ECharts 的 API 和事件处理机制，能够进行自定义的图表交互和事件处理 • 能够结合实际业务场景，利用 ECharts 组件对图表进行定制和扩展 • 具备在使用 ECharts 过程中解决常见问题的能力，如数据格式问题、图表显示异常等
思政目标	• 激发对数据科学的兴趣和热情，增强对数据驱动决策的认识 • 提高团队合作和沟通能力，培养集体意识和协作精神 • 培养创新意识，鼓励自主探索和学习新技术 • 培养职业道德和社会责任感，认识到数据可视化的伦理和社会责任

知识技能储备

2.1　快速上手第一个 ECharts 实例

ECharts 图表入门

下面介绍实现第一个 ECharts 图表，具体步骤如下：

(1) 创建文件夹，重命名为 echartsStudy。

(2) 在 echartsStudy 文件夹下创建两个文件夹，分别命名为 chapter2 和 js，并将下载好的 echarts.js 放入 js 文件夹。

(3) 打开 VsCode 软件后，选择如图 2-1 所示 File 菜单下的“Open Folder”选项。

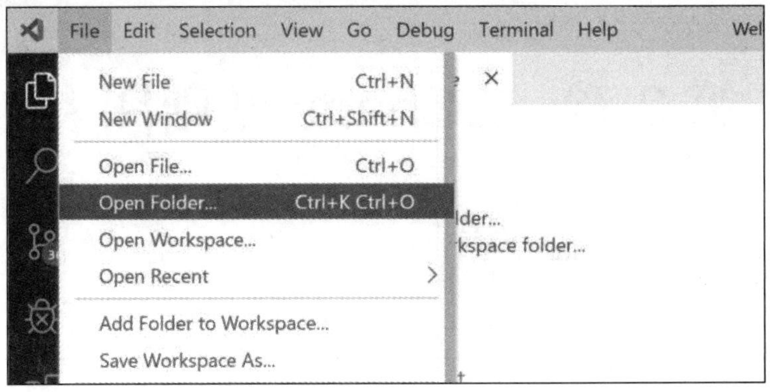

图 2-1　File 菜单

(4) 选择 echartsStudy 目录后，单击如图 2-2 所示的“选择文件夹”按钮，出现如图 2-3 所示的工程视图。

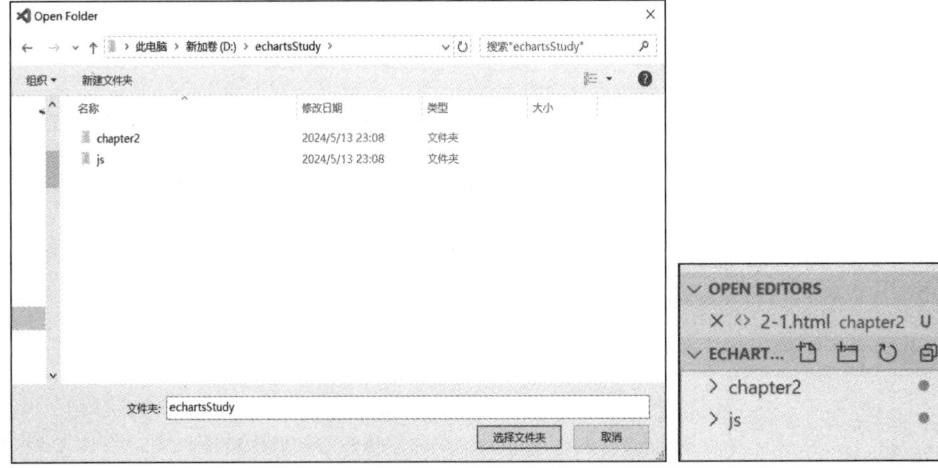

图 2-2　选择文件夹　　图 2-3　工程视图

(5) 选择目录名称“chapter2”，再单击如图 2-4 所示的“New File”图标，输入文件名“2-1.html”。

(6) 在如图 2-5 所示界面的右侧编辑区域输入“html”后，选择“html:5”模板。

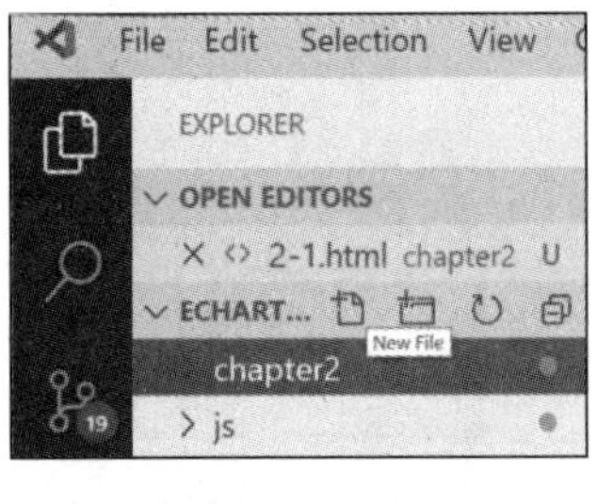

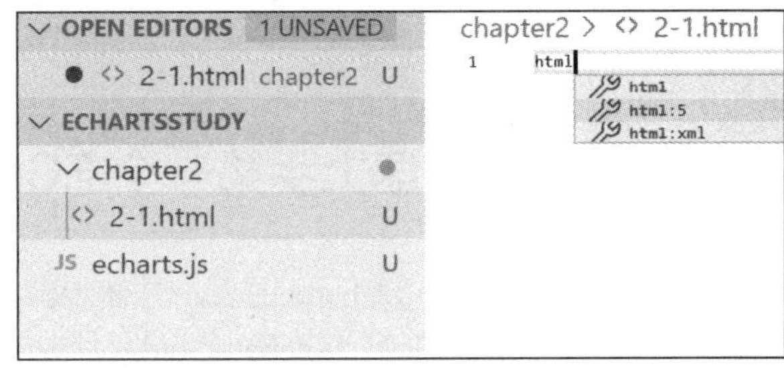

图 2-4　创建文件　　图 2-5　选择模板

(7) 在编辑区域输入文件 2-1.html 中的代码，该代码使用 ECharts 创建了柱状图。

文件 2-1.html

```
<!DOCTYPE html>
<html>
<head>
<meta charset = "UTF-8">
<title>我的第一个程序</title>
<script src = "../js/echarts.js"></script>
</head>
<body>
    //为 ECharts 准备一个具备大小(宽高)的 DOM
    <div id = "main" style = "width:600px; height:400px; "></div>
    <script type = "text/javascript">
    //基于准备好的 DOM，初始化 ECharts 实例
      var mychart = echarts.init(document.getElementById("main"));
    //指定图表的配置项和数据
      var option = {
              //标题组件
              title:{
                  text:"柱状图"
              },
              //通过配置项指定图表的数据和配置
              xAxis:{data:["苹果", "桃子", "香梨", "牛油果", "草莓", "樱桃"]},
              yAxis:{},
              series:[{
                  name:"销量",
                  type:"bar",
                  data:[5, 6, 30, 30, 15, 2]
              }]
      };
      //使用刚指定的配置项和数据显示图表
      mychart.setOption(option);
    </script>
</body>
</html>
```

以上代码表示的 HTML 页面在浏览器中加载时，ECharts 库会被加载，并根据 option 对象的配置，在 id 为 main 的 div 元素中渲染一个柱状图。

(8) 使用浏览器打开 2-1.html 文件，运行结果如图 2-6 所示。

多学一招：当前端页面异常时，可以按下 F12 键开启开发者模式，根据报错的信息排查问题。

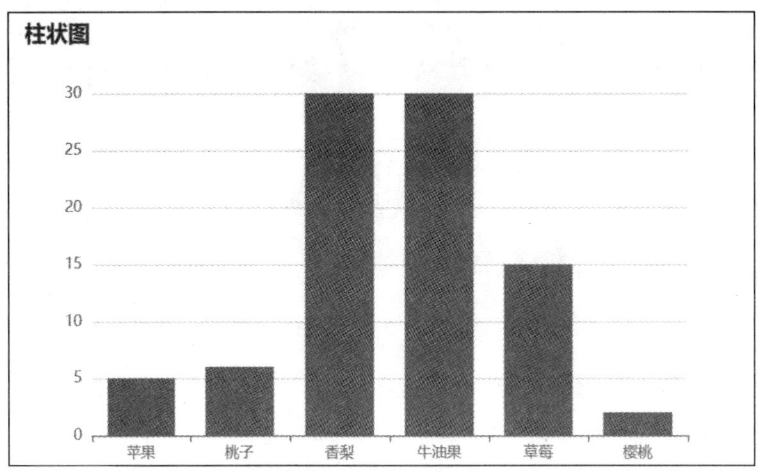

图 2-6 运行结果

从前面第一个入门程序，可以看出 ECharts 图表代码是写在 html 文件中的，其源代码分为以下几个部分：

• 引入 Apache ECharts：需要在代码中引入下载好的 ECharts 文件，具体代码如下：

```
<script src = "../js/echarts.js"></script>
```

• 定义了高和宽的 DOM 容器：通常需要在 HTML 中先定义<div>，并且通过 CSS 使得该 DOM 容器的节点具有宽度和高度，具体代码如下：

```
<div id = "main" style = "width:600px; height:400px; "></div>
```

• 初始化一个 ECharts 实例：可以通过 echarts.init 方法初始化一个实例 ECharts (注意在调用 echarts.init 时需确保 id 前后一致)，具体代码如下。

```
var mychart = echarts.init(document.getElementById("main"));
```

• 配置项：可以通过配置项指定图表的数据和配置，如标题、x/y 坐标轴、图表类型等，具体代码如下：

```
var option = {
        title:{
            text:"柱状图"
        },
         xAxis:{data:["苹果", "桃子", "香梨", "牛油果", "草莓", "樱桃"]},
        yAxis:{},
        series:[{
            name:"销量",
            type:"bar",
            data:[5, 6, 30, 30, 15, 2]
        }]
};
```

以上 title 属性指定了图表的标题。xAxis 属性中定义了 x 方向所有类目名称列表。由于通常情况下 y 方向表示为数值，因此 yAxis 属性没写任何定义，即使用

默认的数值型。series 为数组，可以包含多组对象类型的数据，其中 name 属性定义了该系列名称为 "销量"，用于提示框和图例筛选，type 属性定义图表类型，"bar" 为柱状图，data 属性为该系列的数据。

• 使用刚定义的配置项和数据显示图表，具体代码如下：

```
mychart.setOption(option);
```

该方法执行后，会根据 option 对象的配置，在以上 div 元素中渲染一个柱状图。

2.2 绘制柱状图

2.2.1 绘制标准柱状图

柱状图(或称条形图)是通过柱形的高度来表现数据大小的一种常用图表类型，ECharts 标准柱状图需要做以下基本配置：

(1) xAxis：定义 x 轴的数据，由于横坐标是类目型的，需要在 xAxis 中指定对应的类目名称。

(2) yAxis：定义 y 轴的数据，默认为空；纵坐标默认是数值型的，ECharts 根据 series 中的 data 自动生成对应的坐标范围。

(3) series：定义图表的数据系列，对应对象数组。每个对象包含名称 name、类别 type、数据 data 等常见属性，其中 data 属性为数值型数组，元素的大小决定了柱形的高度；type 属性表示选择图表的类型，若为 bar 则代表柱状图。

【案例 2-1】 文件 2-2-1.html 实现了一个关于销量的柱状图，并对 x 轴、y 轴、数据系列进行基本的配置。

文件 2-2-1.html

```
<!DOCTYPE html>
<html>
<head>
<meta charset="UTF-8">
<title>标准柱状图</title>
//引入 js 文件
<script src="../js/echarts.js"></script>
</head>
<body>
    //为 ECharts 准备一个具备大小(宽高)的 DOM
    <div id="main" style="width: 600px; height: 400px;"></div>
    <script type="text/javascript">
        //创建一个数组
        var x_string = new Array(12);
        var mydata = [3020, 4800, 3600, 6050, 4300, 7200, 4151, 5500, 3020, 4800, 3600, 6050];
```

```
        var data_name = "销量";
        for(var i = 0; i < x_string.length; i++)
        {    x_string[i] = i.toString();
        }
        //将参数指定的内容输出到控制台中，方便调试代码
        console.log(x_string.join(", "));
        //基于准备好的 DOM，初始化 ECharts 实例
        var myechart = echarts.init(document.getElementById("main"));
        //指定图表的配置项和数据
        var option = {
            //主标题内容样式
            title : {
                text : "这是主标题",
                textStyle : {
                    color : "#fff"
                }
            },
        //副标题内容样式
            subtitle : {
                text : "这是副标题",
                subtextStyle : {
                    color:"#bbb"
                }
            },
            legend:{
                show:true   //显示图例
            },
            // x 轴和 y 轴数值定义，通过配置项指定图表的数据和配置
            xAxis:{
                data:x_string
            },
            yAxis:{     },
            series:[ {
                    name:data_name,
                    type:"bar",
                    data:mydata
                }
            ]};
    //使用刚指定的配置项和数据显示图表
```

```
        myechart.setOption(option);
    </script>
</body>
</html>
```

以上代码定义了 x_string 数组，用于存储 x 轴的标签；定义 mydata 为一个数值类型的数组，表示每个柱子的高度；定义 data_name 为一个字符串，表示数据的名称。最后将这些数组配置到 xAxis(x 轴的数据)、series(图表的系列数据)对象的名称和数据中。

使用浏览器运行该文件，结果如图 2-7 所示。

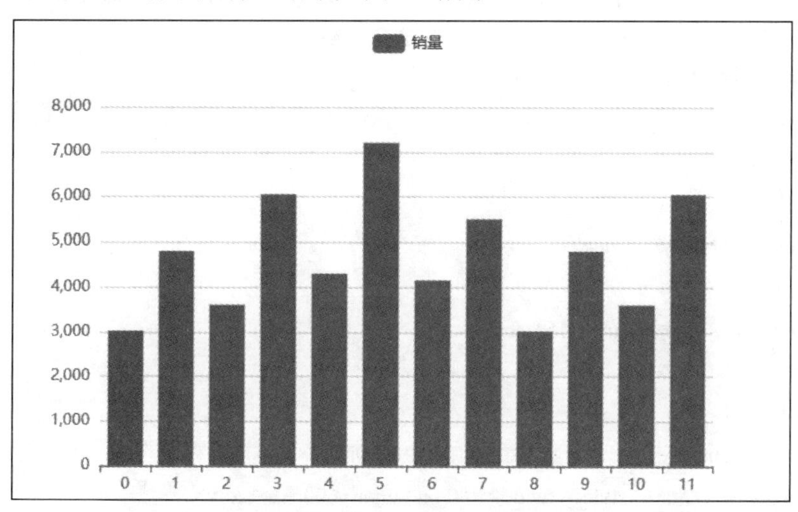

图 2-7　标准柱状图案例运行结果

多学一招：由于实际项目中对图形的个性化要求较多，导致使用的配置项较多，开发者很难记忆所有的配置项，为了方便快速地查询所需配置项内容，需要了解 ECharts 官方文档。

- https://echarts.apache.org/zh/option.html：配置项参考手册。
- https://echarts.apache.org/handbook/zh/get-started/：快速入门的官方地址。
- https://echarts.apache.org/examples/zh/index.html：官方提供的所有实例。

2.2.2　绘制堆叠柱状图

在数据可视化应用场景中，需要显示不同类别或组的数据在某个时间段内的累积情况，可以使用堆叠柱状图(也称为堆积柱状图)来表现。堆叠柱状图就是一个系列的数值“堆叠”在另一个系列上，因而从它们的高度总和就能表达总量的变化，效果如图 2-8 所示。

使用 ECharts 实现堆叠柱状图的方法非常简单，只需要给系列成员对象的 stack 属性设置一个字符串类型的值，拥有同样 stack 值的对象将堆叠在一起。与标准柱状图不同的是：series 属性对应的对象数组中有 2 个及 2 个以上对象，需要堆叠的对象填入相同的 stack 属性。

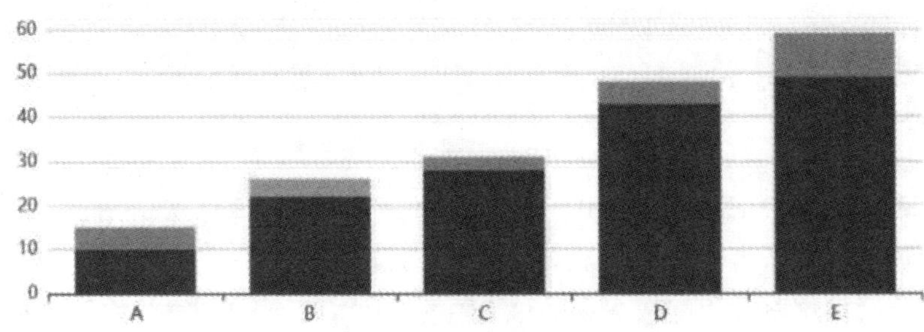

图 2-8　堆叠柱状图效果

【案例 2-2】 文件 2-2-2.html 实现了广告类型的数据的堆叠显示。

文件 2-2-2.html

```
<!DOCTYPE html>
<html>
<head>
<meta charset = "UTF-8">
<title>堆叠柱状图</title>
//引入 js 文件
<script src = "../js/echarts.js"></script>
</head>
<body>
    //为 ECharts 准备一个具备大小(宽高)的 DOM
    <div id = "main" style = "width: 800px; height: 600px"></div>
    <script type = "text/javascript">
        //基于准备好的 DOM，初始化 ECharts 实例
        var mychart = echarts.init(document.getElementById("main"));
        var len = 7;
        var xdata = new Array(len);
        for (var i = 0; i < len; i++) {
            xdata[i] = "星期"+(i+1).toString();
        }
        console.log(xdata.join(", "));
        var mydata = new Array(4);
        for (var i = 0; i < 4; i++) {
            mydata[i] = new Array(len);
            for (var j = 0; j < len; j++) {
                mydata[i][j] = Math.random() * 100;
            }
        }
        //指定图表的配置项和数据
        var option = {
```

```
            title : {//标题组件
                text : "堆积条形图"
            },
            //x 轴和 y 轴数据系列
            xAxis : {
                data : xdata
            },
            yAxis : {},
            series : [ {
                    name : "直接访问",
                    type : "bar",
                    data : mydata[0]
                }, {
                    type : "bar",
                    data : mydata[1],
                    name : "视频广告",
                    stack : "广告"
                }, {
                    type : "bar",
                    data : mydata[2],
                    name : "联盟广告",
                    stack : "广告"
                }, {
                    name:"游戏",
                    type : "bar",
                    data : mydata[3],
                } ],
            legend:{   //图例组件
                data:legend_array
            }
        };
        var legend_array = new Array(option.series.length);   //定义图例数组
        for(var i = 0; i < legend_array.length; i++){
            legend_array[i] = option.series[i].name;
        }
        mychart.setOption(option);   //使用刚指定的配置项和数据显示图表
    </script>
</body>
</html>
```

以上代码首先初始化了一个 ECharts 实例，并创建 x 轴和 y 轴的数据。然后定义一个名为“option”的对象，该对象包含了图表的配置项，如标题、x 轴、y 轴、数据系列 series 等，并在 series 数据中给视频广告和联盟广告填入相同的 stack 属性。接着定义一个图例数组并将配置项中的系列名称添加进来。最后使用指定的配置项和数据显示图表。

使用浏览器打开后，效果如图 2-9 所示。

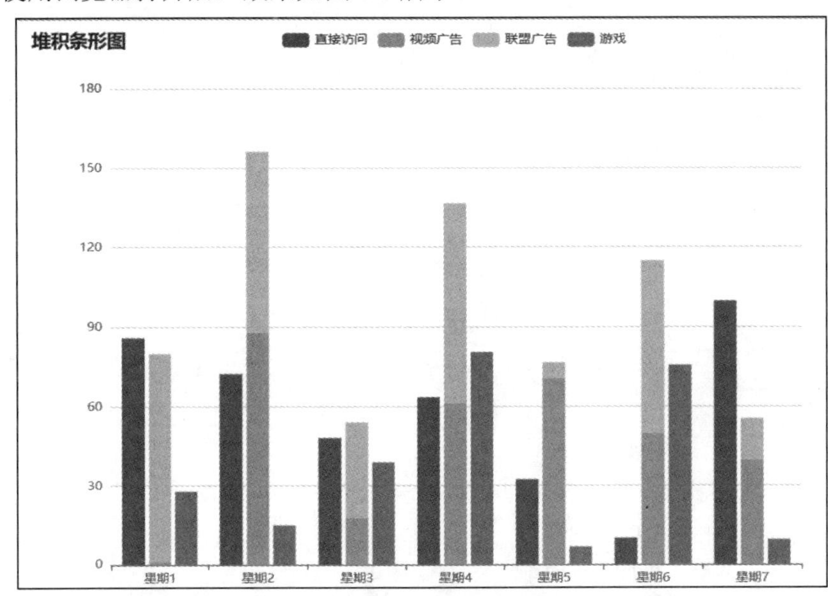

图 2-9　堆叠柱状图案例运行效果

2.2.3　绘制标准条形图

标准条形图即横向柱状图，与普通条形图差异在于 x 轴和 y 轴的类别不同，即 y 轴变成了分类轴，x 轴变成了数值轴。条形图适用于比较不同类别数据，尤其适用于类别较多或类别名称较长的场景。因此，在 ECharts 中只需要在普通柱状图的基础上，把 x 轴和 y 轴调换设置即可实现该效果。

【案例 2-3】文件 2-2-3.html 实现了关于 6 个省份销售增长率的标准条形图，需要创建一个包含 6 个省份名称的数组 ydata，再生成一个包含 6 个随机数的数组 mydata，每个随机数的范围为 0～50。定义一个名为“option”的对象，用于存储图表的配置项和数据，其中包括标题、x 轴、y 轴和系列等配置。

文件 2-2-3.html

```
<!DOCTYPE html>
<html>
<head>
<meta charset = "UTF-8">
<title>标准条形图即横向柱状图</title>
</head>
```

```
<script src = "../js/echarts.js"></script>
<body>
<div id = "main" style = "width:600px; height:400px"></div>
<script type = "text/javascript">
    var mychart = echarts.init(document.getElementById("main"));
    var len = 6;
    var ydata = new Array(len);
    var i = 0;
    while(i < len){
        ydata[i] = String.fromCharCode(("A".charCodeAt(0)+i))+"省";
        i++;
    }
    console.log(ydata.join(", "));
    var mydata = new Array(6);
    i = 0;
    while(i < len){
        mydata[i] = Math.random()*50; // 生成一个 0～50 的整型随机数
        i++;
    }
    var option = {                          //指定图表的配置项和数据
            title:{                         //标题组件
                text:"标准条形图"
            },
            xAxis:{                         //通过配置项指定图表的数据和配置
                type:"value"
            },
            yAxis:{
                type:"category",
                data:ydata
            },
            series:[ {
                    type:"bar",
                    data:mydata
                } ]
    };
//使用刚指定的配置项和数据显示图表
    mychart.setOption(option);
</script>
</body>
</html>
```

以上代码实现的条形图中，y 轴为分类轴，x 轴为数值轴。可以看出条形图通过长度差异更直观地反映数值的差异，而柱状图则是通过高度来展示这种差异。以上条形图配置的运行效果如图 2-10 所示。

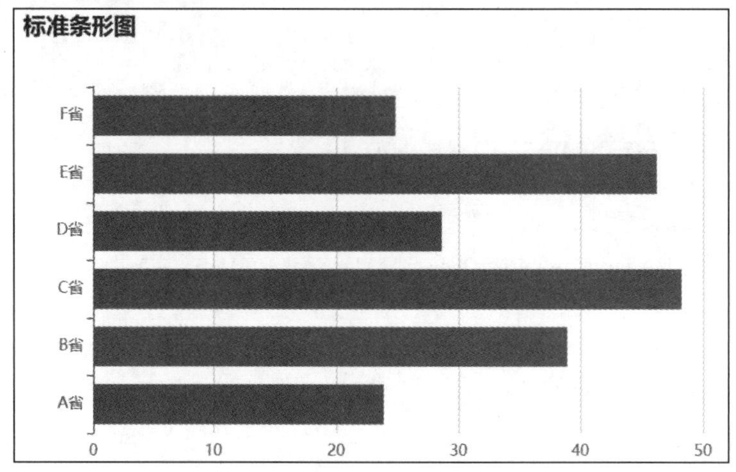

图 2-10 标准条形图运行效果

2.2.4 绘制瀑布图

瀑布图通常用于展示数据点之间的演变过程，当拆解项较多时，瀑布图相比于饼图能更清晰地展示每一项对总量的贡献或影响。瀑布图是一种特殊类型的柱状图，差异在于从第 2 列数据开始首尾相接且“悬空”的效果，如图 2-11 所示。Apache ECharts 中并没有单独的瀑布图类型，但是可以使用堆叠柱状图间接实现该效果。

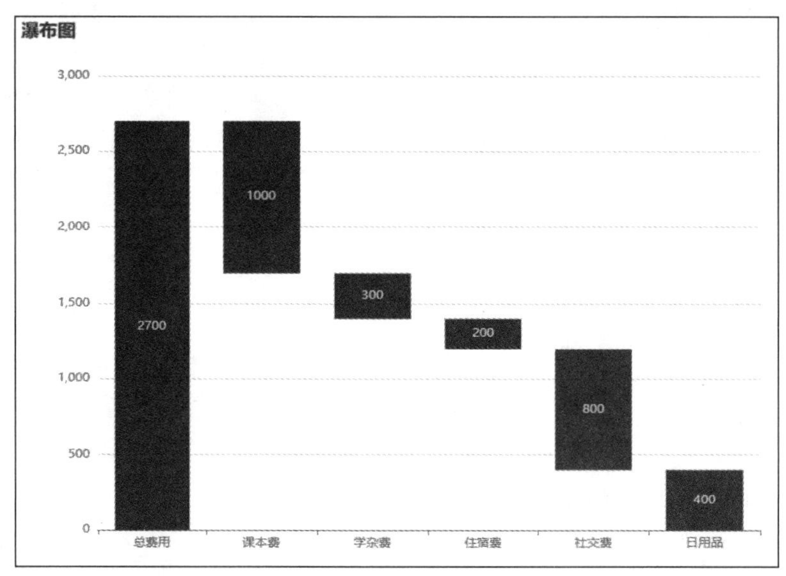

图 2-11 瀑布图效果

【案例 2-4】文件 2-2-4.html 使用堆叠柱状图的两个系列实现了瀑布图效果：

第 1 个系列是不可交互的透明系列，用来实现“悬空”的柱状图效果；第 2 个系列真正用来表示展示数据大小。

文件 2-2-4.html

```
<!DOCTYPE html>
<html>
<head>
<meta charset = "UTF-8">
<title>瀑布图</title>
<script src = "../js/echarts.js"></script>
</head>
<body>
    <div id = "main" style = "width: 800px; height: 600px" />
    <script type = "text/javascript">
        var mychart = echarts.init(document.getElementById("main"));
        var xdata = ["总费用", "课本费", "学杂费", "住宿费", "社交费", "日用品"];
        var data1 = [ 0, 1700, 1400, 1200, 400, 0 ]
        var data2 = [ 2700, 1000, 300, 200, 800, 400 ]
        //指定图表的配置项和数据
        var option = {
            title : {
                text : "瀑布图"
            },
            xAxis : {
                data : xdata
            },
            yAxis : {},
            series : [ {
                name : "辅助",
                type : "bar",
                stack : "总量",
                itemStyle:{
                  normal:{
                barBorderWidth:5,
                color:"rgba(10, 10, 10, 0)"
                // color:"rgba(10, 10, 10, 0.5)"
                }
              },
            data : data1
            }, {
                /* 上方的柱子 */
```

```
                name:"生活费",
                type:"bar",
                stack:"总量",
                 itemStyle:{
                    normal:{
                        color:"rgba(200, 0, 0, 100)",
                        label:{
                            show:true,
                            position:"inside"
                            }
                }
                    },
                data:data2
            }
            ],
        };
        var legend_array = new Array(2);
        legend_array[0] = option.series[0].name;
        legend_array[1] = option.series[1].name;
        //使用刚指定的配置项和数据显示图表
        mychart.setOption(option);
    </script>
</body>
</html>
```

以上这两个系列的数据，定义为 data1 和 data2，分别表示下方的柱子和上方的柱子，它们都是包含 6 个元素的数组，分别表示每个费用项的具体数值。因此可以得知，瀑布图实质上就是堆叠柱状图，但是区别在于 series 对象数组的大小为 2，其中第一个对象的 ItemStyle 的颜色为透明色。

多学一招： 柱状图的样式配置，如透明度、颜色等详见 ECharts 官方文档(https://echarts.apache.org/zh/option.html)的配置项手册。

2.3 绘制折线图

2.3.1 绘制标准折线图

折线图主要用来展示数据项随着时间推移的趋势或变化，折线图效果如图 2-12 所示。类似于柱状图，折线图横坐标是类目型(category)，纵坐标是数值型(value)，但是需要配置系列对象的 type 为 line，表示该图形为折线图。另外一个非常重要的参数是 smooth，即选择是否平滑。

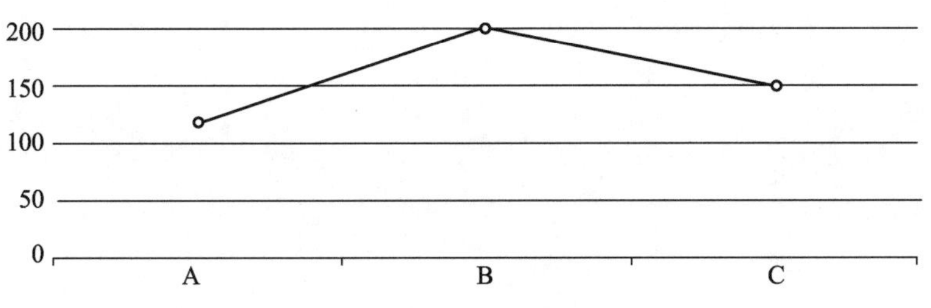

图 2-12　折线图效果

【案例 2-5】 文件 2-3-1.html 使用 ECharts 库创建一个平滑效果的折线图。创建变量 data_name、mydata 和 x_label，分别用于存储图表的数据名称，数据值和横坐标标签；配置 option 的对象，用于存储图表的配置项和数据；该折线图只有一个数据系列，类型为 line，平滑曲线为 true，名称为 data_name[0]，数据为 mydata。

文件 2-3-1.html

```
<!DOCTYPE html>
<html>
<head>
<meta charset = "UTF-8">
<title>我的折线图</title>
<script src = "../js/echarts.js"></script>
</head>
<body>
    <div id = "main" style = "width: 600px; height: 400px;" />
    <script type = "text/javascript">
        var mychart = echarts.init(document.getElementById("main"));
        var data_name = ["人流量"] ;
        var mydata = new Array(7);
        var i = 0;
        while (i < 7) {                  //生成一个 0～10000 的整型随机数
            mydata[i] = Math.random() *10000;
            i++;
        }
        //横坐标标签
        var x_label = [ "1 月", "2 月", "3 月", "4 月", "5 月", "6 月", "7 月" ];
        var option = {                  //指定图表的配置项和数据
                title : {               //标题组件
                text : "标准折线图",
                textStyle : {
                    color : "red",
                    fontSize : 10
                },
                x : "center"            //标题居中
```

```
            },
            /*提示框 坐标轴触发*/
            tooltip : {
                trigger : "axis"
            },
            legend : {              //图例
                data : data_name,
                x : "right"
            },
            xAxis : {
                type : "category",
                data : x_label
            },
            yAxis : {},
            series : [{
                type : "line",      //折线图类型
                smooth : true,      //true 为平滑曲线，false 为直线
                name:data_name[0],
                data : mydata
            }]
        };
        mychart.setOption(option);
    </script>
</body>
</html>
```

在上述代码的 option 对象中，设置了标题、提示框、图例、x 轴和 y 轴等组件；类似于柱状图，其 x 轴类型为类别，数据为 x_label，y 轴类型为空；而提示框触发方式为坐标轴触发，图例显示为“人流量”。该代码运行效果如图 2-13 所示。

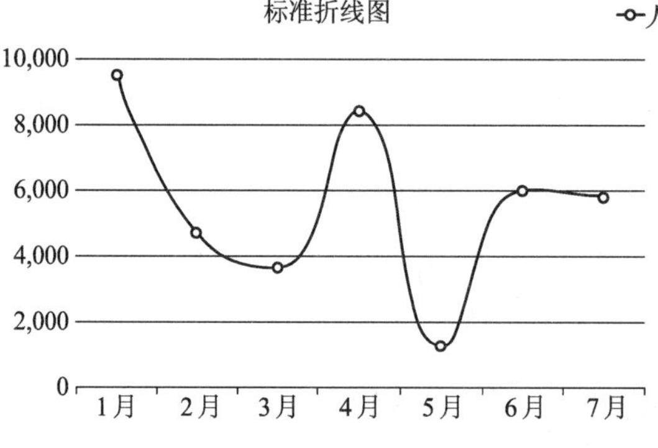

图 2-13　平滑效果的折线图

多学一招：以上项目使用了图例和提示框，详情可参考 3.3 节的内容，同时了解 ECharts 官方文档的配置项参考手册。

2.3.2　绘制堆积折线图和堆积面积图

与堆叠柱状图类似，可以将折线图堆叠在一起，堆积折线图效果如图 2-14 所示。堆叠折线图是建立在普通折线图基础上，用系列的 stack 设置需要堆叠在一起的数据。

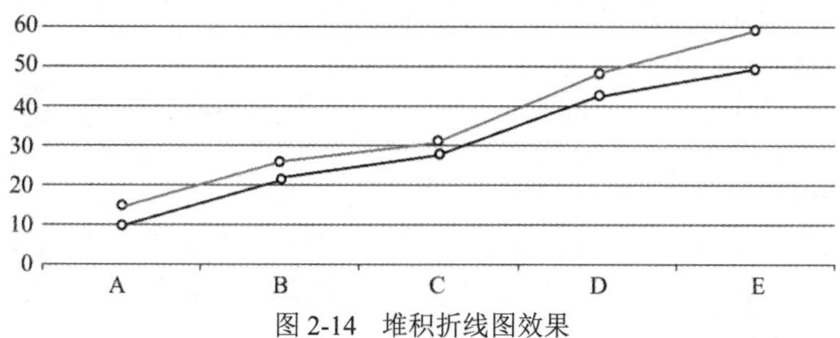

图 2-14　堆积折线图效果

从对应配置项代码中可以看出与堆叠柱状图相比区别仅仅在于 type 属性不同，具体代码如下：

```
option = {
  xAxis: {data: ["A", "B", "C", "D", "E"]},
  yAxis: {},
  series: [{
      data: [10, 22, 28, 43, 49],
      type: "line",
      stack: "x"
    },{
      data: [5, 4, 3, 5, 10],
      type: "line",
      stack: "x"
    } ]
};
```

对于堆叠折线图而言，一般建议使用区域填充色以表明堆叠的情况，这样就成为堆积面积图，堆积面积图是在普通折线图基础上加上 stack 属性和 areaStyle 属性构成的。

【案例 2-6】 文件 2-3-2.html 中实现了一个堆积面积图，具有以下功能：展示多个类别的数据变化趋势、显示数据累积效果；用户将鼠标悬停在图表上时，可以查看每个数据点的详细信息；允许用户选择显示或隐藏某个类别的数据；提供一些导航工具，如数据视图(dataView)和数据缩放(dataZoom)，方便用户查看和分析特定时间段或数据范围。

文件 2-3-2.html

```
<!DOCTYPE html>
<html>
<head>
```

```
<meta charset = "UTF-8">
<title>堆积面积图</title>
<script src = "../js/echarts.js"></script>
</head>
<body>
    <div id = "main" style = "width: 600px; height: 400px;"></div>
    <script type = "text/javascript">
        var mychart = echarts.init(document.getElementById("main"));
        var size = 5;
        var len = 7;
        var data_name = [ "烤箱", "扫地机器人", "咖啡机", "净水机", "料理机" ];
        var x_label = [ "一月", "二月", "三月", "四月", "五月", "六月", "七月" ];
        var mydata = new Array();
        var i = 0;
        while (i < size) {
            var j = 0;
            mydata[i] = new Array(len);
            while (j < len) {
                mydata[i][j] = Math.random() * 500;
                j++;
            }
            i++;
        }
        //指定图表的配置项和数据
        var option = {
                title : {   //标题组件
                    text : "堆积面积图",
                    textStyle : {
                        color : "blue",
                        fontSize : 20,
                        fontWeight : "bold"
                    }
                },
                xAxis : {
                    type : "category",
                    data : x_label
                },
                yAxis : {},
                series : [ {
                    type : "line",
```

```
        stack : "总量",
        data : mydata[0],
        name : data_name[0],
        color : "rgb(0, 0, 0)",
        areaStyle:{}
    }, {
        type : "line",
        stack : "总量",
        data : mydata[1],
        name : data_name[1],
        color : "blue",
        areaStyle:{}
    }, {
        type : "line",
        stack : "总量",
        data : mydata[2],
        name : data_name[2],
        color : "green",
        areaStyle:{}
    }, {
        type : "line",
        stack : "总量",
        data : mydata[3],
        name : data_name[3],
        color : "red",
        areaStyle:{}
    }, {
        type : "line",
        stack : "总量",
        data : mydata[4],
        name : data_name[4],
        color : "yellow",
        areaStyle:{}
    } ],
    /* 提示框 */
    tooltip:{
        trigger:"axis"
    },
    /*图例*/
    legend:{
```

```
                        data:data_name,
                        left:160,
                        top:3
                    },
                    toolbox:{/*工具箱*/
                        orient:"vertical",
                        feature:{/*工具配置项*/
                            saveAsImage:{/*保存为图片*/
                            },
                            restore:{/*配置项还原*/
                            },
                            dataView:{/*数据视图工具*/
                            },
                            dataZoom:{/*放大*/
                            }
                        }
                    }
                };
            mychart.setOption(option);   //使用刚指定的配置项和数据显示图表
        </script>
    </body>
</html>
```

以上代码定义的变量 size 表示数据系列的数量，即堆积面积图中的类别数量；变量 len 表示每个数据系列中的数据点数量，即堆积面积图中的时间轴长度；data_name 为一个包含 5 个字符串的数组，表示每个数据系列的类别名称；x_label 为一个包含 7 个字符串的数组，表示时间轴上的标签；mydata 为一个二维数组，mydata[i][j]表示第 i 个数据系列在第 j 个时间点的数据值，在 while 循环中 Math.random()方法生成 mydata 的每个元素。

最终运行效果如图 2-15 所示。

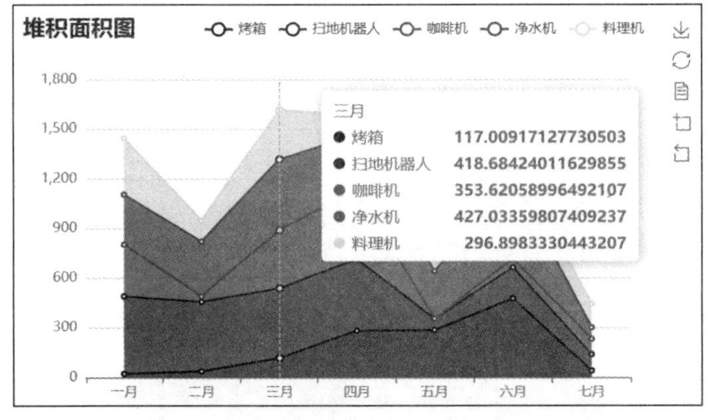

图 2-15 堆积面积图运行结果

多学一招：以上项目使用了工具箱，详情可参考 3.3 节的内容和 ECharts 官方文档的配置项手册。

2.3.3　绘制阶梯图

阶梯线图又称方波图，主要使用水平线和垂直线连接两个数据点，如图 2-16 所示。而普通折线图则直接将两个点连接起来。阶梯线图能够很好地表达数据的突变。在 ECharts 中，阶梯线图类似于折线图，其类型 type 为 line，但是需要指定 step 属性为 true 显示成阶梯线图。step 属性也可以设置成其他 3 种类型，即 start、middle 和 end，分别表示在当前点、当前点与下个点的中间点、下个点这 3 处出现过渡。

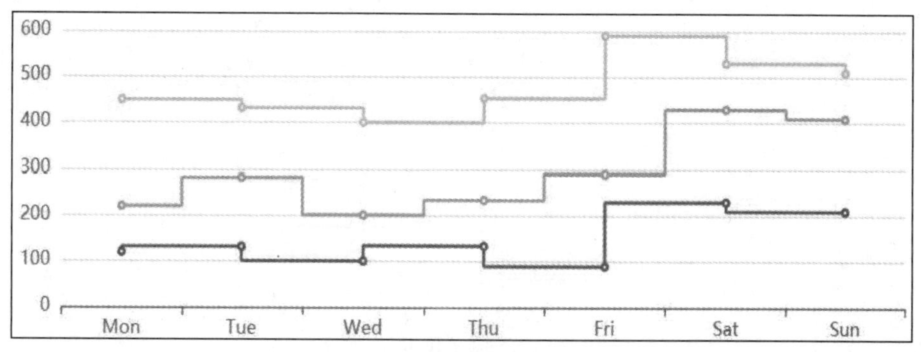

图 2-16　阶梯图效果

【案例 2-7】文件 2-3-3.html 实现阶梯图效果，保证数据序列的类型 type 为 line 且 step 属性为 true，以显示成阶梯线图。

文件 2-3-3.html

```
<!DOCTYPE html>
<html>
<head>
<meta charset = "UTF-8">
<title>标准阶梯图</title>
<script src = "../js/echarts.js"></script>
</head>
<body>
    <div id = "main" style = "width: 600px; height: 400px;"></div>
    <script type = "text/javascript">
        var x_string = new Array(12);        //定义数组
        var mydata = [3020, 4800, 3600, 6050, 4300, 7200, 4151, 5500, 3020, 4800, 3600, 6050];
        var data_name = "销量";                    //定义变量
        for(var i = 0; i < x_string.length; i++)
        {
            x_string[i] = i.toString();
```

```
            }
            console.log(x_string.join(", "));
            var myechart = echarts.init(document.getElementById("main"));
            var option = {
            title : {
                show:true,
                text :"阶梯图",
                textStyle : {
                    color : "#ff0000",
                    fontWeight:"bold",          //字体加粗
                    fontSize:30                 //字号
                },
                x:"center"                      //标题在 x 轴中间
            },
            legend:{                            //图例在 x 轴右侧
                show:true,
                x:"right"
            },
            xAxis:{                             //添加 x 轴
                data:x_string
            },
            yAxis:{
            },
            series:[
                 {
                    name:data_name,
                    type:"line",                //生成 line 类型图表
                    step:true,
                    data:mydata
                 }
            ]
        };
        myechart.setOption(option);
    </script>
</body>
</html>
```

以上代码在 series 属性中定义一个数据系列，类型为 line，并设置 step 属性为 true，表示生成阶梯折线图；最后将 option 对象应用到 myechart 实例上，从而渲染出阶梯折线图，最终运行效果如图 2-17 所示。

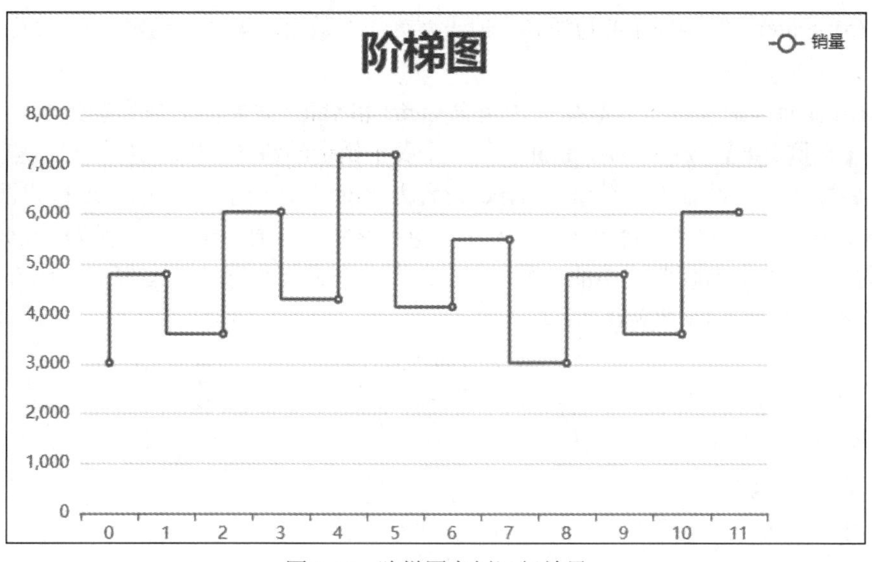

图 2-17　阶梯图案例运行效果

2.4　绘 制 饼 图

2.4.1　绘制标准饼图

饼图主要用于表现不同类目的数据在总和中的占比，其每个扇形的弧度表示数据数量的比例，饼图效果如图 2-18 所示。

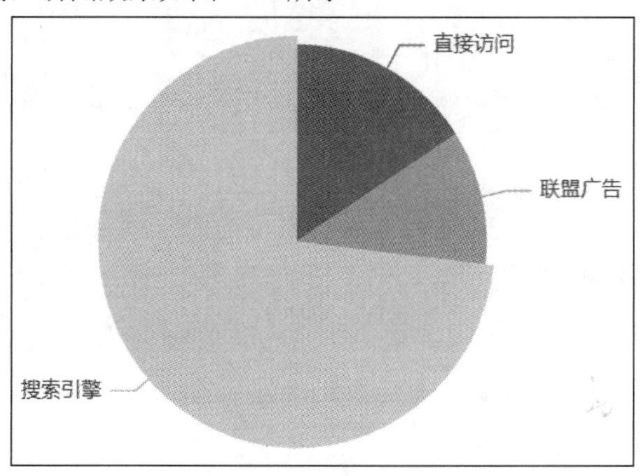

图 2-18　饼图效果

饼图的配置和折线图、柱状图有显著区别，它不再需要配置坐标轴，而是直接配置数据系列的以下属性：

(1) name：数据系列的名称。

(2) type：数据系列的图表类型，设置为 pie 表示饼图。

(3) label：标签的格式器。

(4) radius：饼图的半径，表示饼图相对容器大小的百分比。

(5) center：饼图的中心位置，一般设置为一个数组，第一个元素为 x 坐标，第二个元素为 y 坐标。

(6) data：配置项系列数据，由包含 name 和 value 属性的对象数组组成。

【案例 2-8】 文件 2-4-1.html 实现一个关于销量的饼图，代码定义一个数据系列，属性 name 设置为“销量”；type 设置为“pie”，表示饼图；label 格式器为“{b}:{d}%”，表示显示每个扇区的标签，包括名称和占比；radius 设置为“40%”，表示饼图的大小为容器大小的 40%；center 第一个元素为圆心的 x 坐标(50%)，第二个元素为圆心的 y 坐标(60%)。

文件 2-4-1.html

```
<!DOCTYPE html>
<html>
<head>
<meta charset = "UTF-8">
<title>我的第一个饼图</title>
<script src = "../js/echarts.js"></script>
</head>
<body>
    <div id = "main" style = "width:600px; height:400px; "></div>
    <script type = "text/javascript">
      var mychart = echarts.init(document.getElementById("main"));
      var name_data = ["衬衫", "羊毛衫", "裤子", "袜子", "高跟鞋", "帽子"];
      var value_data = new Array();
      var mydata = new Array();
      for( myname in name_data){
              mydata.push({value:Math.random()*100, name:name_data[myname]});
      }
      console.log(mydata[1]);
      var option = {
              title:{
                  text:"饼图"
              },
              series:[{
                  name:"销量",
                  type:"pie",   //图表类型为 pie 即为饼图
                  label:{
                      formatter:"{b}:{d}%"
                  },
                  radius:"40%",
                  center:["50%", "60%"],
                  data:mydata
```

```
            }],
            tooltip:{
                trigger:"item",
                 formatter:"{a}<br/>{b}:{c}({d}%)"
            },
            toolbox:{
                x:"center",
                show:true,
                feature:{
                    saveAsImage:{show:true},              //保存为图像
                    restore:{show:true},                  //重置
                    dataView:{show:true, readOnly:false}  //数据试图
                }
            }
        };
        mychart.setOption(option);
    </script>
</body>
</html>
```

以上代码的提示框 tooltip 的 formatter 属性，其模板变量取值有以下几种：{a}表示系列名；{b}为数据名；{c}表示数据值；{d}表示百分比；{@xxx}表示数据中名为“xxx”的维度值，如{@product}表示名为“product”的维度值；{@[n]}表示数据中维度 n 的值，如{@[3]}表示维度 3 的值，从 0 开始计数。

使用浏览器打开文件 2-4-1.html，最终运行效果如图 2-19 所示。

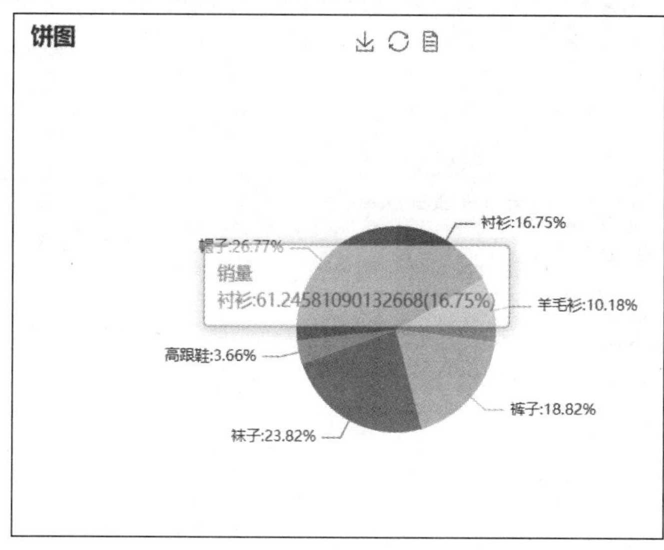

图 2-19 饼图案例运行效果

多学一招：以上项目使用了提示框，参考 3.3 节内容或了解 ECharts 官方文档的配置项手册。

2.4.2 绘制圆环图

在 ECharts 中，饼图的半径可以是一个数值或者字符串之外，还可以是一个包含两个元素的数组，当它是一个数组时，它的前一项表示内半径，后一项表示外半径，这样就形成了一个圆环图，因此圆环图与普通饼图的区别在于圆环图有内外两个半径。

【案例 2-9】 文件 2-4-2.html 实现一个圆环图，定义了一个数据系列，包含以下属性：

(1) name：数据系列的名称，这里设置为“销量”。

(2) type：数据系列的图表类型，这里设置为“pie”，表示饼图。

(3) data：数据系列的数据，这里使用之前定义的 mydata 数组作为数据源。

(4) radius：饼图的半径，这里设置为一个数组，第一个元素为内半径(30%)，第二个元素为外半径(45%)。

(5) center：饼图的中心位置，这里设置为一个数组，第一个元素为 x 坐标(50%)，第二个元素为 y 坐标(60%)。

文件 2-4-2.html

```
<!DOCTYPE html>
<html>
<head>
<meta charset = "UTF-8">
<title>圆环饼图</title>
<script src = "../js/echarts.js"></script>
</head>
<body>
    <div id = "main" style = "width:600px; height:400px; "></div>
    <script type = "text/javascript">
       var mychart = echarts.init(document.getElementById("main"));
       var name_data = ["衬衫", "羊毛衫", "裤子", "袜子", "高跟鞋", "帽子"];
       var value_data = new Array();
       var mydata = new Array();
       for( myname in name_data){
           /*  注意 js 中的 for in 变量为下标*/
           mydata.push({value:Math.random()*100, name:name_data[myname]});
       }
       console.log(mydata[1]);  //在浏览器的控制台中打印信息
       var option = {
               title:{
```

```
                text:"圆环饼图"
            },
            series:[{
                name:"销量",
                type:"pie",
                data:mydata,
                radius:["30%", "45%"],
                center:["50%", "60%"]
            }]
        };
        mychart.setOption(option);
    </script>
</body>
</html>
```

以上代码通过这些属性的配置，生成一个具有指定名称、类型、数据、半径和中心位置的饼图数据系列，最终运行效果如图 2-20 所示。

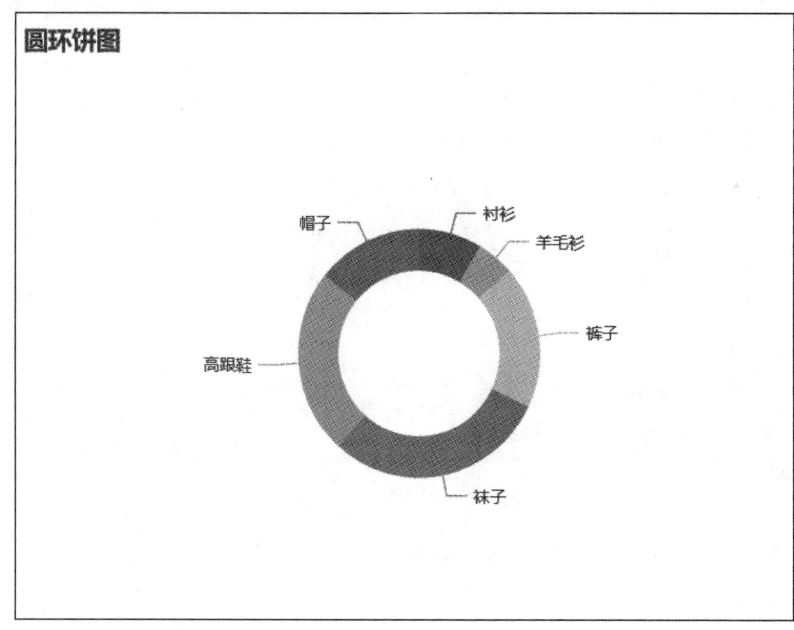

图 2-20　圆环图案例运行效果

2.4.3　绘制嵌套饼图

嵌套饼图效果如图 2-21 所示，它同时嵌套了普通饼图和多个圆环饼图，这些圆环需要保证在同一个圆心上。因此使用 ECharts 实现该效果需要准备多个数据系列，使每个数据系列 center 属性相同而半径不同。

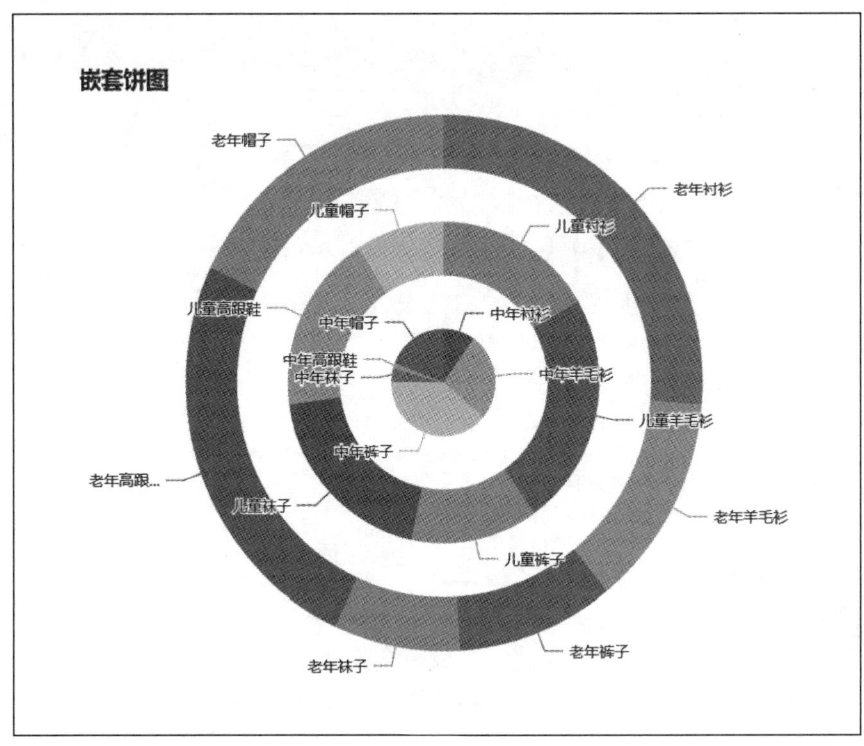

图 2-21 嵌套饼图效果

【案例 2-10】 文件 2-4-3.html 实现图 2-21 所示的嵌套饼图，为了提高代码的复用，该文件通过 js 生成该嵌套饼图的配置信息。

文件 2-4-3.html

```
<!DOCTYPE html>
<html>
<head>
<meta charset = "UTF-8">
<title>嵌套饼图</title>
<script src = "../js/echarts.js"></script>
</head>
<body>
    <div id = "main" style = "width:600px; height:500px; "></div>
    <script type = "text/javascript">
        var mychart = echarts.init(document.getElementById("main"));
        var name_data = ["衬衫", "羊毛衫", "裤子", "袜子", "高跟鞋", "帽子"];
        var name_series = ["中年", "儿童", "老年"];
        var mydata = new Array(3);
        for(var i = 0; i < 3; i++){
            mydata[i] = [];
            for( myname in name_data){ /*注意 js 中的 for in 变量为下标*/
```

```
                //对数组 mydata 中的第 i 个元素进行操作，向其中添加一个新的对象
                mydata[i].push({value:Math.random()*100,name:name_series[i]+name_data[myname]});
            }
        }
        console.log(mydata[0]);
        var config = new Array(3);
        for(var i = 0; i < 3; i++){                          // for 循环生成 3 个饼图的配置
            var rad1 = ((100/6)*2*i).toString()+"%";         //计算 rad1 的值
            var rad2 = ((100/6)*(2*i+1)).toString()+"%";     //计算 rad2 的值
        console.log(rad1+":"+rad2);
        config[i] = {
                name:name_series[i],
                type:"pie",                                  //嵌套饼图图表类型依然为 pie
                radius:[rad1, rad2],
                data:mydata[i]
            }
        }
        var option = {
            title:{
                text:"嵌套饼图"
            },
            series:config
        };
        mychart.setOption(option);
    </script>
</body>
</html>
```

在以上代码中首先定义了两个数组 name_data 和 name_series，分别存储数据名称和系列名称；然后通过循环遍历这两个数组，生成了对象数组 mydata，其每个元素都是一个包含 value 和 name 属性的对象；接着通过循环遍历 name_series 数组，生成一个 config 数组，其每个元素都是一个包含 name、type、radius 和 data 属性的对象，用于配置嵌套饼图的样式和数据。

2.5　绘制散点图和气泡图

2.5.1　绘制散点图

散点图也是一种常见的图表类型，这种图形由许多“点”组成，这些点用来

表示数据在坐标系中的位置。散点图的基本配置选项包括 x 轴、y 轴和数据系列：

(1) xAxis：表示 x 轴的配置，空对象表示使用默认配置。

(2) yAxis：表示 y 轴的配置，空对象表示使用默认配置。

(3) series：表示数据系列。对于数据系列的 symbolSize 属性表示散点的大小；data 属性表示散点的坐标数据，每个元素是一个包含 x 和 y 值的数组；type 属性为“scatter”，表示图表类型为散点图；属性 symbol 表示散点图中数据“点”的形状，ECharts 内置形状包括圆形、矩形、三角形、菱形、圆角矩形、大头针形和箭头形，分别对应“circle”“rect”“triangle”“diamond”“roundRect”“pin”“arrow”。

【案例 2-11】 文件 2-5-1.html 实现了一个包含标记点和标记线的散点图。

文件 2-5-1.html

```
<!DOCTYPE html>
<html>
<head>
<meta charset = "UTF-8">
<title>散点图</title>
<script src = "../js/echarts.js"></script>
</head>
<body>
    <div id = "main" style = "width: 600px; height: 400px;"></div>
    <script type="text/javascript">
        var mydata = [ ];
        var data_name = "销量";
        for (var i = 0; i <12; i++) {
            mydata.push([Math.random()*20, Math.random()*20]);
        }
        var myechart = echarts.init(document.getElementById("main"));
        var option = {                          //定义图表的配置项和数据
            title : {
                show : true,                    //是否显示标题
                text : "散点图",                //标题文本内容
                textStyle : {                   //标题文本样式
                    color : "#00ffff"           //标题文本颜色为青色
                }
            },
            legend : {                          //图例配置
                show : true
            },
            xAxis : {                           //x 轴配置，是否包含 0 刻度
                scale:false
```

```
    },
    yAxis : {scale:false},              //y 轴配置
    series : [ {                        //系列列表
        name : data_name,
        type : "scatter",               //图表类型为散点图
        data : mydata,
        markPoint : {                   //标记点
            symbolSize:20,
            data : [ {
                type : "max",
                name : "最大",
                symbol : "arrow",       //最大点形状为箭头
                itemStyle : {
                    color : "red"
                }
            },
            {
               type:"min",
               name:"最小",
               symbol:"diamond",        //最小点形状为菱形
               itemStyle:{
                   color:"yellow"       //最小点颜色为黄色
               }
            }]
        },
        markLine:{                      //标记线
            data:[{
                type:"average",         //标记平均值线
                name:"平均值",
                lineStyle:{             //线的样式
                    color:"darkred",
                    type:"dotted",
                    width:2
                }
            }]
        }
    } ]
};
```

```
        myechart.setOption(option);
    </script>
</body>
</html>
```

以上代码在 series 中定义一个散点图类型的数据系列，散点图的名称是在 js 中定义的 data_name 变量，数据为 mydata 数组；定义标记点和标记线的配置项，分别用于显示最大值、最小值和平均值。运行结果如图 2-22 所示。

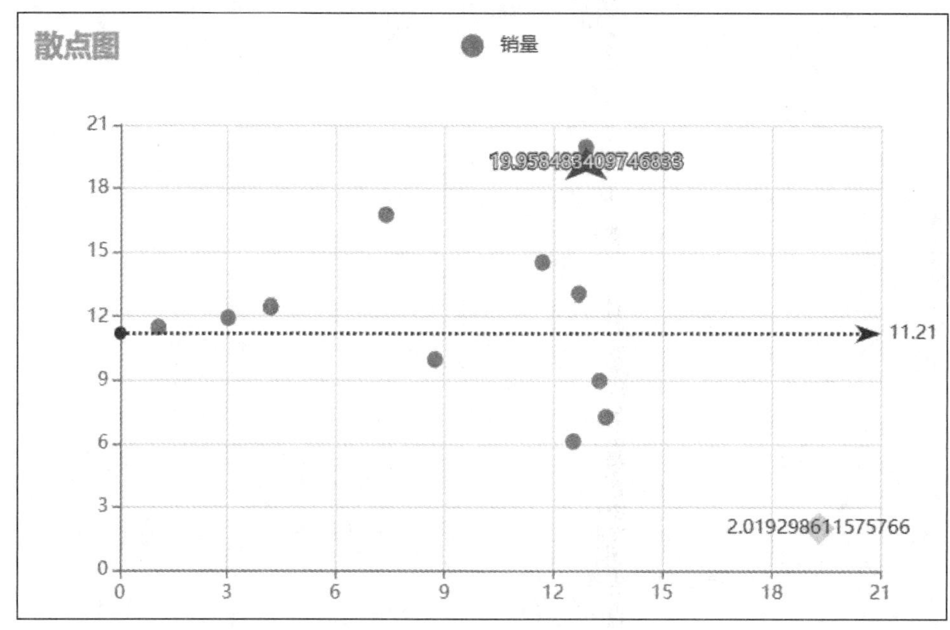

图 2-22 散点图案例运行效果

2.5.2 绘制气泡图

气泡图是一种数据可视化方法，是一种扩展的散点图，它利用点的大小来表达额外的数据维度，为分析和理解多变量关系提供一个有力的视觉效果。它在普通散点图数据基础上增加了一个额外的维度，并且使用属性 symbolSize 控制气泡大小。

【案例 2-12】 文件 2-5-2.html 实现了一个气泡图，包含两个数据系列：汽车销量和玩具销量。

文件 2-5-2.html

```
<!DOCTYPE html>
<html>
<head>
<meta charset="UTF-8">
<title>气泡散点图</title>
<script src="../js/echarts.js"></script>
```

```
</head>
<body>
    <div id="main" style="width: 600px; height: 400px;"></div>
    <script type="text/javascript">
    // 初始化两个空数组用于存储数据
        var mydata1 = [];
        var mydata2 = [];
        // 定义数据名称数组
        var data_name = ["汽车销量", "玩具销量"];
        // 循环生成 12 组随机数据
        for (var i = 0; i < 12; i++) {
            mydata1.push([ Math.random() * 20, Math.random() * 20]);
            mydata2.push([ Math.random() * 20, Math.random() * 20 ]);
        }
        console.log(mydata1);
        var myechart = echarts.init(document.getElementById("main"));
         // 定义图表的配置项
        var option = {
            title : {
                show : true,
                text : "气泡图",
                textStyle : {
                    color : "#00ffff"
                }
            },
            xAxis : {                                    //x 轴配置项，是否包含 0 刻度
                scale : false
            },
            yAxis : {                                    // y 轴配置项
                scale : false
            },
            legend : {
                show : true
            },
            series : [ {                                 //第 1 个系列配置项
                name : data_name[0],                     //系列名称
                type : "scatter",                        //图表类型为散点图
                data : mydata1,                          //数据内容
                color:"green",                           //数据点颜色为绿色
```

```
                symbolSize:function(value){          //气泡大小函数
                    return Math.random()*30;         //返回 0～30 的随机大小
                }
            }, {                                     //第 2 个系列配置项
                name : data_name[1],                 //必配
                type : "scatter",                    //必配
                data : mydata2,                      //必配
                symbol:"arrow",                      //数据点形状为箭头
                symbolSize:function(value){          //必配
                    return Math.round(value[1]);
                }
            } ]
        };
        myechart.setOption(option);
    </script>
</body>
</html>
```

在以上代码中，第一个数据系列的图表类型设置为散点图(scatter)，数据内容为 mydata1，数据点颜色为绿色，symbolSize 属性对应的气泡大小函数返回值为 0～30 的随机数；设置了第二个数据系列的图表类型为散点图，数据内容为 mydata2，数据点形状为箭头，气泡大小函数返回值为将 mydata2 [1]经过四舍五入后的整数值。

在浏览器中运行以上代码，最终运行效果如图 2-23 所示。

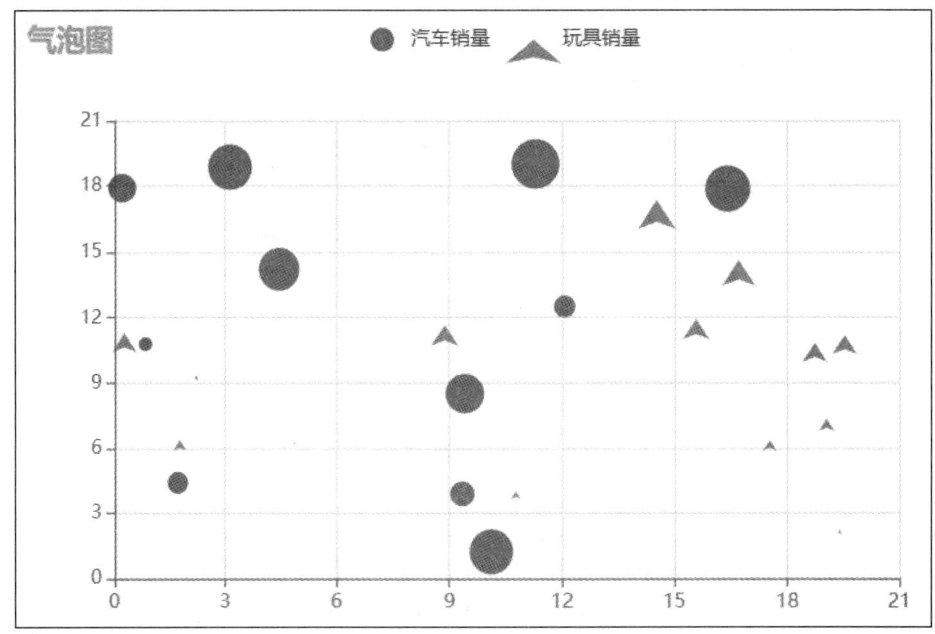

图 2-23　气泡图案例运行效果

2.6 实战演练：电商销售数据分析网页

1. 需求分析

需求分析如下：

序号	需求类别	需求描述
1	图表展示	展示 6 个不同品类的月销售数据，每个品类用堆积面积图表示 展示 6 个品类的年度销售总额和总销售额 展示 6 个品类的年度销售数量占比 展示 6 个品类的平均价格
2	交互功能	实现图表的数据提示，提升用户体验 提供一个工具箱，允许用户保存图表为图片

2. 规划设计

填写如下规划设计表：

<table>
<tr><td colspan="4">WBS 表</td></tr>
<tr><td colspan="4">项目基本情况</td></tr>
<tr><td>项目名称</td><td>电商销售数据分析网页</td><td>任务编号</td><td></td></tr>
<tr><td>姓　　名</td><td></td><td>班　　级</td><td></td></tr>
<tr><td colspan="4">工作分解</td></tr>
<tr><td>工作任务</td><td colspan="2">包含活动</td><td>备注</td></tr>
<tr><td rowspan="2">1. 环境搭建</td><td colspan="2">1.1 下载 js 包</td><td></td></tr>
<tr><td colspan="2">1.2 拷贝 ECharts 包至本地文件夹</td><td></td></tr>
<tr><td rowspan="5">2. 项目实现</td><td colspan="2">2.1 数据准备</td><td></td></tr>
<tr><td colspan="2">2.2 绘制分月销售数量堆积面积图</td><td></td></tr>
<tr><td colspan="2">2.3 绘制销售金额瀑布图</td><td></td></tr>
<tr><td colspan="2">2.4 绘制全年销售饼图</td><td></td></tr>
<tr><td colspan="2">2.5 绘制价格对比柱状图</td><td></td></tr>
</table>

3. 搭建环境

(1) 从官网 https://echarts.apache.org/ 下载完整 js 包。

(2) 本地新建文件夹，并复制下载的 ECharts 包到 js 文件夹。

4. 项目实现

1) 数据准备

需要初始化数据，生成堆积面积图的数据系列，并计算总销售额。这些数据将用于后续的图表绘制，具体代码如下：

```
var myseries1 = []
var seriesName = []
var price_avg = [10, 8, 4, 3, 8, 5]
var salesCount = []
var totalSalesCount = 0;
var xdata = ["一月", "二月", "三月", "四月", "五月", "六月", "七月", "八月", "九月", "十月",
"十一月", "十二月"];
for (var i = 0; i < 6; i++) {  // 生成 6 个品类的数据
        seriesName.push("品类" + String.fromCharCode("A".charCodeAt(0) + i))
        var myobj = {
                name: seriesName[i],
                type: "line",
                stack: "总量",
                data: []
        }
        for (var j = 0; j < 12; j++) {  // 为每个品类生成 12 个月份的数据
                var data_temp = Math.round(Math.random() * 300) + 50
                myobj.data.push(data_temp)
        }
         myseries1.push(myobj)  // 将当前品类的数据对象添加到折线图的数据系列中
          // 计算销售金额
          salesCount.push(myobj.data.reduce(
                (accumulator, currentValue) => accumulator + currentValue, 0) * price_avg[i])
}
// 计算总销售额
totalSalesCount = salesCount.reduce((accumulator, currentValue) => accumulator +
currentValue, 0)
```

这部分代码中变量 myseries1 包含了堆积面积图的数据，seriesName 将代表品类名称，price_avg 是每个品类的平均价格，salesCount 将用于存储每个品类的销售总额，totalSalesCount 将存储所有品类的总销售额，xdata 是 x 轴的月份数据。代码使用局部变量 myobj 生成了 6 个不同的数据系列，每个系列代表一个品类。每个品类的数据 data_temp 是通过随机数生成的，代表该品类在每个月的销售数量。同时计算了每个品类的销售金额，并存储在 salesCount 数组中。最后使用 reduce 方法计算所有品类的总销售额，并将结果存储在 totalSalesCount 变量中。

2) 绘制分月销售数量堆积面积图

初始化第 1 个 ECharts 实例，并配置堆积面积图的选项，然后使用这些选项在页面上显示堆积面积图，具体格式如下：

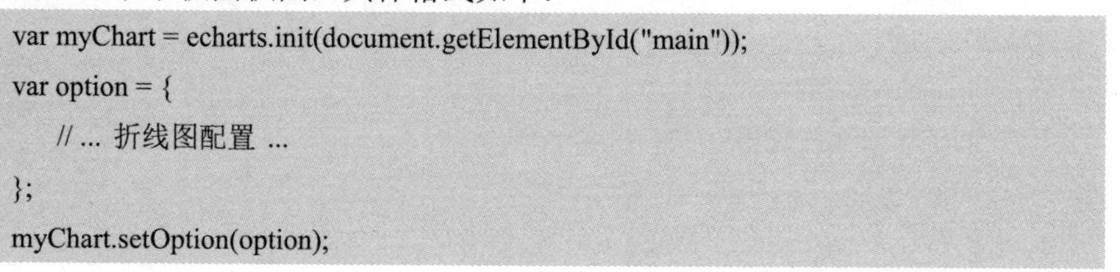

```
var myChart = echarts.init(document.getElementById("main"));
var option = {
    // ... 折线图配置 ...
};
myChart.setOption(option);
```

3) 绘制销售金额瀑布图

初始化第 2 个 ECharts 实例，并配置瀑布图的选项，包括计算瀑布图所需的数据，然后使用这些选项在页面上显示瀑布图，具体格式如下：

```
var myChart2 = echarts.init(document.getElementById("main2"));
var xdata2 = ["总费用"].concat(seriesName);
// ... 瀑布图数据计算 ...
var option2 = {
    // ... 瀑布图配置 ...
};
myChart2.setOption(option2);
```

4) 绘制全年销售饼图

初始化第 3 个 ECharts 实例，并配置饼图的选项，包括计算饼图所需的数据，然后使用这些选项在页面上显示饼图，具体格式如下：

```
var data3 = []
// ... 饼图数据计算 ...
var myChart3 = echarts.init(document.getElementById("main3"));
var option3 = {
    // ... 饼图配置 ...
};
myChart3.setOption(option3);
```

5) 绘制价格对比柱状图

初始化第 4 个 ECharts 实例，并配置柱状图的选项，然后使用这些选项在页面上显示柱状图，具体格式如下：

```
var myChart4 = echarts.init(document.getElementById("main4"));
var option4 = {
    // ... 柱状图配置 ...
};
myChart4.setOption(option4);
```

6) 最终 html 文件实现

为了生成最终图表，合并和实现以上 5 个模块代码，生成文件 salesData.html。

文件 salesData.html

```
<!DOCTYPE html>
<html>
<head>
    <meta charset="utf-8">
    <title>ECharts 电商销售数据柱状图</title>
    <script src="../../js/echarts.js"></script>
    <style>
        .mycontainer {
            display: flex;
            justify-content: space-around;
            width: 1100px;
            margin: 30px auto;
        }
        .mydiv {
            width: 500px;
            height: 400px;
            border: 1px solid gray;
        }
    </style>
</head>
<body>
    <h1 style="text-align: center;">电商销售数据</h1>
    <div class="mycontainer">
        <div id="main" class="mydiv"></div>
        <div id="main2" class="mydiv"></div>
    </div>
    <div class="mycontainer">
        <div id="main3" class="mydiv"></div>
        <div id="main4" class="mydiv"></div>
    </div>
    <script type="text/javascript">
        var myseries1 = []
        var seriesName = []
        var price_avg = [10, 8, 4, 3, 8, 5]
        var salesCount = []
        var totalSalesCount = 0;
        var xdata = ["一月", "二月", "三月", "四月", "五月", "六月", "七月", "八月", "九月",
        "十月", "十一月", "十二月"];
```

```
for (var i = 0; i < 6; i++)
{
    seriesName.push("品类" + String.fromCharCode("A".charCodeAt(0) + i))
    var myobj = {
        name: seriesName[i],
        type: "line",
        stack: "总量",
        areaStyle:"true",
        data: []
    }
    for (var j = 0; j < 12; j++)
    {
        var data_temp = Math.round(Math.random() * 300) + 50
        myobj.data.push(data_temp)
    }
    myseries1.push(myobj)
    // 计算销售金额
    salesCount.push(myobj.data.reduce(
        (accumulator, currentValue) => accumulator + currentValue, 0) * price_avg[i])
}
// 计算总销售额
totalSalesCount = salesCount.reduce((accumulator, currentValue) => accumulator +
currentValue, 0)
// 绘制
var myChart = echarts.init(document.getElementById("main"));
var option = {
    title: {
        text: "分月销售数量",
        left: "center"
    },
    tooltip: {
        trigger: "axis",
        axisPointer: {
            type: "shadow"
        }
    },
    legend: {
        top: "20"
    },
```

```
    grid: {
        left: "3%",
        right: "4%",
        bottom: "3%",
        containLabel: true
    },
    toolbox: {
        feature: {
            saveAsImage: {}
        }
    },
    xAxis: {
        type: "category",
        boundaryGap: false,
        data: xdata
    },
    yAxis: {
        type: "value"
    },
    series: myseries1
};
myChart.setOption(option);
// 销售金额瀑布图
var myChart2 = echarts.init(document.getElementById("main2"));
var xdata2 = ["总费用"].concat(seriesName);
// 下面的柱子
var data1 = new Array(xdata.length).fill(0)
data1[0] = 0
// 上面的柱子
var data2 = [totalSalesCount].concat(salesCount)
// 计算后续元素
for (var i = seriesName.length; i > 1; i--)
{
    // 每个元素是前面所有元素的总和
    data1[i - 1] = data1[i] + data2[i];
}
console.log(data1)
//指定图表的配置项和数据
var option2 = {
```

```
        title: {   //标题组件
            text: "销售金额",
            left: "center"
        },
        xAxis: {
            data: xdata2
        },
        yAxis: {},
        series: [{  //下方的柱子
            type: "bar",
            stack: "总量",
            itemStyle: {
                normal: {
                    barBorderWidth: 5,
                    color: "rgba(10, 10, 10, 0)"
                }
            },
            data: data1
        }, {
            /* 上方的柱子 */
            type: "bar",
            stack: "总量",
            itemStyle: {
                normal: {
                    color: "rgba(200, 0, 0, 100)",
                    label: {  //内部填充文字 name
                        show: true,
                        position: "inside"
                    }
                }
            },
            data: data2
        }],
    };
    var legend_array = new Array(2);
    legend_array[0] = option.series[0].name;
    legend_array[1] = option.series[1].name;
    myChart2.setOption(option2);
    //全年销售饼图
```

```
var data3 = []
for (var i = 0; i < seriesName.length; i++)
{
    var total = 0
    for (var j = 0; j < 12; j++)
    {
        total = total + myseries1[i].data[j]
    }
    data3.push({ value: total, name: seriesName[i] })
}
var myChart3 = echarts.init(document.getElementById("main3"));
var option3 = {
    title: {
        text: "全年销售数量",
        left: "center"
    },
    legend: {
        top: "20px"
    },
    series: [
        {
            name: "各系列全年销售占比",
            type: "pie",
            data: data3
        }
    ]
};
myChart3.setOption(option3);
// 价格对比柱状图
var myChart4 = echarts.init(document.getElementById("main4"));
var option4 = {
    title: {
        text: "销售价格",
        left: "center"
    },
    xAxis: {
        data: seriesName
    },
    yAxis: {},
```

```
                series: [
                    {
                        name: "各系列价格",
                        type: "bar",
                        data: price_avg
                    }
                ]
            };
            myChart4.setOption(option4);
        </script>
    </body>
</html>
```

以上代码通过生成堆积面积图、瀑布图、饼图和柱状图，提供了不同维度的销售数据分析，包括时间序列分析、金额分布、总销售额占比和价格比较。这些图表可以帮助用户更好地理解销售数据的各个方面。

7) 运行结果

将文件 salesData.html 使用浏览器打开，运行结果如图 2-24 所示。

电商销售数据

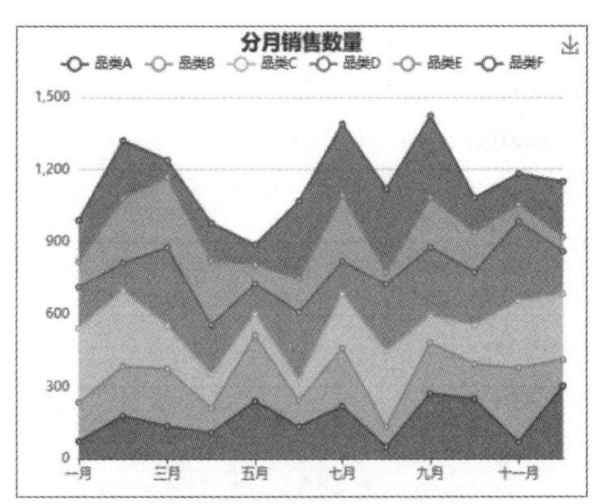

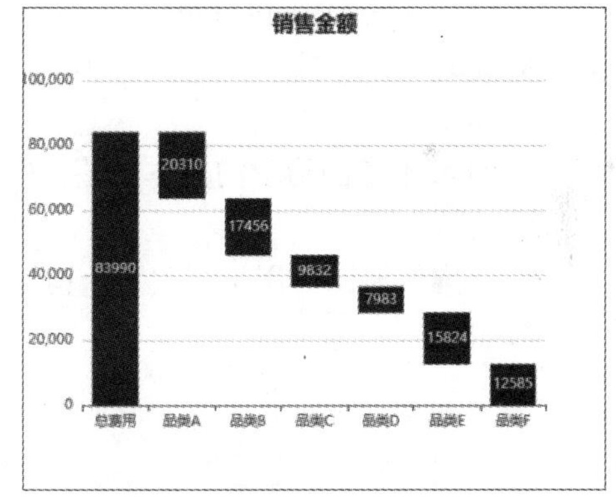

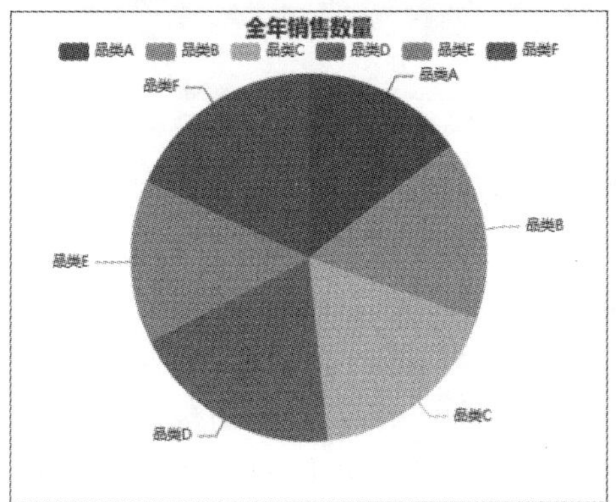

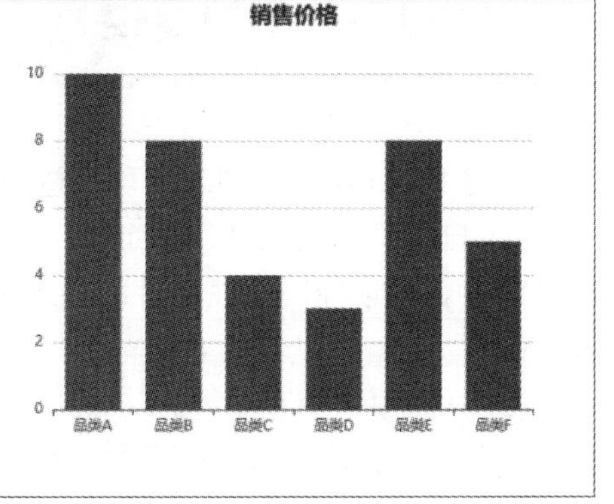

图 2-24　电商销售数据运行结果

通过运行结果可以看出：以上分月销售数量堆积面积图显示了 6 个不同品类(A～F)在一年中每个月的销售数量；销售金额瀑布图展示了 6 个品类的销售金额以及总销售额；全年销售饼图显示代表每一个品类的总销售数

量占比；价格对比柱状图展示了 6 个品类的平均价格。

强化练习

习题答案

1. symbol 为圆形应该选(　　)。

A. circle　　B. rect
C. roundRect　　D. triangle

2. barWidth 是柱状图的(　　)。

A. 高度　　B. 宽度
C. 柱子宽度　　D. 柱子高度

3. label. position 取值有(　　)。

A. top　　B. left
C. right　　D. bottom

4. (　　)不是饼图的 type。

A. pie　　B. bar
C. line　　D. scatter

5. label 不是图形上的文本标签。(　　)

A. 正确　　B. 错误

6. 标准气泡图可用于观察 3 个指标的关系。(　　)

A. 正确　　B. 错误

进一步学习建议

学习 ECharts 常用图表，可以进一步学习以下内容，以提高技能和理解：

(1) 自定义图表组件。ECharts 允许自定义图表的标题、图例、工具箱、提示框等组件。可以控制这些组件的显示位置、内容和样式，以对图表的细节进行精细控制。

(2) 学习复杂交互功能。ECharts 提供了丰富的交互功能，如数据区域缩放、数据点 hover、联动、事件触发等。建议深入学习这些交互功能的实现原理和应用场景，以便在数据可视化中更好地满足用户需求。

考核评价

考核评价表				
姓名			班　　级	
学号			考评时间	
评价主题及总分		评价内容及分数		评分
1	知识考核(30)	阐述 ECharts 常用图表(柱状图、折线图、饼图、散点图)特点和适用场景(10 分)		
		掌握 ECharts 气泡图的实现原理、功能和使用方法(10 分)		
		熟悉 ECharts 的配置项和参数设置，根据需求进行合理配置(10 分)		
2	技能考核(40)	具备业务需求分析、功能设计，编码及测试的综合能力(10 分)		
		能够按时完成开发任务(20 分)		
		根据给定的数据集选择合适的图表类型，并利用 ECharts 实现数据的可视化，具备实际操作能力(10 分)		
3	思政考核(30)	对数据驱动决策的认识程度，评估数据思维(10 分)		
		具备良好的团队协作精神，能与团队成员有效沟通(10 分)		
		自主探索和学习新技术，职业道德和社会责任感，促进全面发展(10 分)		
评语：				汇总：

第 3 章　ECharts 高级图表及组件

学习目标

目标类型	目 标 描 述
知识目标	• 掌握仪表盘的基本概念和构成，包括指针、刻度盘、标签等元素的配置方法，能够使用仪表盘展示数据 • 掌握漏斗图的基本概念和构成，包括漏斗的形状、数据表示方式、边框等，能够使用漏斗图展示数据 • 掌握 ECharts 常用组件的功能及配置
技能目标	• 掌握定制仪表盘的方法，如设置半径大小、起始角度、刻度线样式等，利用仪表盘展示数据 • 掌握定制漏斗图、金字塔图的方法，如设置颜色、高亮、标签等，利用漏斗图、金字塔图展示数据 • 掌握配置常用组件来增强图表的交互性和可读性的方法，如设置标题文本、图例位置、工具箱功能等
思政目标	• 培养坚定的自我驱动力，形成良好的学习习惯，并不断提升自学能力，以主动的态度持续学习和掌握新的知识和技能 • 培养严谨的逻辑思维和细致的工作态度，在编写代码时，始终坚守细心与耐心，严格规范地书写每一个配置项，确保代码的准确性和可靠性 • 强化问题解决能力，能够独立分析并有效解决在配置展示图过程中遇到的各类问题 • 在学会高级图表的配置后，将其应用在我们的学习过程中，如可以使用仪表盘定制学习目标，用颜色表示不同的学习进度，使用漏斗图展示学习的转化率，通过图表数据显示分析学习中的瓶颈及问题，从而定制更贴合自身的学习计划

知识技能储备

3.1　绘制仪表盘

3.1.1　绘制单个仪表图形

仪表盘(Gauge)也称拨号图表或速度表图，通常用于展示一个或多个关键性能

指标的当前状态，如完成率、进度、速度等，使用户能够快速了解核心数据所在的范围和状态。仪表盘效果图如图 3-1 所示。

图 3-1　仪表盘效果

图 3-1 所示的仪表盘图形，其配置项如下：

```
option = {
  series: [
    {
      name: "Pressure",
      type: "gauge",
      data: [
        {
          value: 50,
          name: "temp"
        }
      ]
    }
  ]
};
```

由以上可知仪表盘图没有 x 轴和 y 轴，图形类别 type 属性对应为 gauge，数据包含 name 和 value 两个属性。ECharts 提供了丰富的配置选项，允许用户根据实际的数据和场景自定义仪表盘的各种元素，如指针、刻度、颜色等，也可以调整仪表盘的大小、位置、起始角度等参数，以满足不同的展示需求。其他常见属性如下：

(1) center：表示中心(圆心)坐标，值为数组，其第一项是横坐标，第二项是纵坐标，如 center: [400, 300]。

(2) radius：仪表盘半径，可以是相对于容器高或宽中较小项一半的百分比(如高宽为[600, 400]，可设置半径百分比为 50%，此时半径 = (400/2) × 50%)，也可以是绝对的数值。

(3) startAngle：仪表盘起始角度。

(4) endAngle：仪表盘结束角度。

(5) clockwise：表示仪表盘刻度是否按顺时针增长。

(6) min：最小的数据值，映射到 minAngle。

(7) max：最大的数据值，映射到 maxAngle。

(8) splitNumber：仪表盘刻度的分割段数。

(9) axisLine：仪表盘轴线相关配置。

【案例 3-1】 文件 3-1-1.html 实现了一个关于速度的仪表盘，对仪表盘的半径、刻度、速度等进行基本配置。

文件 3-1-1.html

```
<!DOCTYPE html>
<html>
<head>
<meta charset="UTF-8">
<title>仪表盘</title>
<script src="../js/echarts.js"></script>
</head>
<body>
    <div id="main" style="width:600px; height:400px; "></div>
    <script type="text/javascript">
        var mychart=echarts.init(document.getElementById("main"));
        var option={
                title:{
                    text: "仪表盘",
                    x:"center",
                    y:15,
                    textStyle:{
                        fontFamily:"黑体",
                        fontSize:20,
                        color:"blue"
                    }
                },
                series:[{
                    name: "速度",            //必选
                    //表盘内部的标题，内容同以上的 name
                    title:{
                      offsetCenter:[0, "-30%"]
                    },
                    type: "gauge",           //必配项
                    data:[{                  //必配项
```

```
        name:"速度",
        value:50
    }],
    min:0,
    max:200,                    //默认 100
    splitNumber:10,             //可以不配置，默认 10
    axisLine:{                  //仪表轮廓线
        show:true,
        lineStyle:{
            //各个区间颜色
            color:[[0.2, "rgba(255, 0, 0, 1)"],
                   [0.8, "rgba(255, 255, 0, 1)"],
                   [1.0, "rgba(0, 255, 0, 1)"]],
            opacity:0.8,        //透明度默认 1
            width:30,
            shadowBlur:20,      //阴影面积，即发光效果
            shadowColar:"#000fff"
        }
    },
    radius:"90%",
    center:["50%", "60%"],
    startAngle:225,
    endAngle:-45,
    clockWise:true,             //默认就是 true
    splitLine:{                 //大的分割线
        length:30,
        lineStyle:{
            color:"#000eee",
            opacity:1,
            width:2,
            shadowBlur:10,
            shadowColor:"#0fff"
        }
    },
    axisTick:{                  //小刻度线
        splitNumber:5,          //默认就是 5
        length:8,
        lineStyle:{
            color:"#0eee",
```

```
                        shadowBlur:10,
                        shadowColor:"#0fff",
                        width:1.5,
                    }
                },
                //刻度标签
                axisLabel:{
                    distance:30,            //与刻度的距离
                    color:"blue",
                    fontSize:15,
                },
                //指针
                pointer:{
                    length: "50%",
                    width:15
                },
                //指针样式
                itemStyle:{
                    color:"auto",           //取数值所在区间颜色
                    //边线宽及颜色
                    borderWidth:3,
                    borderColor:"#000",
                    shadowBlur:10,          //发光效果
                    shadowColor:"orange"
                },
                emphasis:{},
                 //下方的数值
                detail:{
                    offsetCenter:[0, "60%"],
                    color:"black",
                    fontSize:30,
                    formatter:"{value}km/h"
                }
            }]
    };
//间隔 1.5 s 更新一次数据
  setInterval(function(){
     //改变显示的值，结果为保留两位小数
     option.series[0].data[0].value=(Math.random()*120).toFixed(2);
```

```
        /* 第 2 个参数是否不立即更新图表，默认为 false，即同步立即更新。如果为 true，
        则会在下一个 animation frame 中才更新图表 */
            mychart.setOption(option, true);
        }, 1500);
    </script>
</body>
</html>
```

以上代码设置当前速度范围为 0～200，默认值为 50；仪表盘有 10 个刻度线，分为 3 个颜色区间，分别为红色、黄色和绿色；指针的颜色根据数值所在的区间自动变化；下方显示当前速度值，单位为 km/h；每隔 1.5 s 更新一次速度值，保留两位小数。

使用浏览器运行该文件，结果如图 3-2 所示。

图 3-2　仪表盘案例运行效果

3.1.2　绘制多个仪表图形

单仪表盘一般用于一个特定的指标或目标的监控、分析和预警，实现方式相对比较简单。当需要同时监控和比较多个关键指标时，从多个维度综合评估，跟踪多个参数开展数据关联分析，需要多仪表盘构建一个信息丰富的仪表板。

【案例 3-2】 文件 3-1-2.html 实现一个汽车仪表盘面板，对仪表盘的半径、位置、刻度、显示弧度等进行基本配置。

文件 3-1-2.html

```
<!DOCTYPE html>
<html>
<head>
<meta charset="UTF-8">
<title>多仪表盘</title>
<script src="../js/echarts.js"></script>
</head>
```

```
<body>
    <div id="main" style="width:1200px; height:600px; "></div>
    <script type="text/javascript">
        var mychart = echarts.init(document.getElementById("main"));
        var option = {
            title: {
                text: "多仪表盘-汽车仪表盘",
                x: "center",
                y: 55,
                textStyle: {
                    fontFamily: "黑体",
                    fontSize: 40,
                    color: "#000066",
                }
            },
            series: [
                {                                           //中间仪表盘
                    name: "车速",                           //必选
                    //表盘内部的标题，内容同以上的 name
                    title: {
                        offsetCenter: [0, "-30%"]
                    },
                    type: "gauge",                          //必配项
                    data: [{                                //必配项
                        name: "速度",
                        value: 50
                    }],
                    min: 0,
                    max: 200,                               //默认 100
                    splitNumber: 10,                        //可以不配置，默认为 10
                    axisLine: {                             //仪表轮廓线
                        show: true,
                        lineStyle: {
                            color: [                        //各个区间颜色
                                [0.3, "green"],             // 0%～30%
                                [0.7, "rgba(255, 255, 0, 1)"],  // 30%～70%
                                [1.0, "rgba(255, 0, 0, 1)"]],   // 70%～100%
                            opacity: 0.8,                   //透明度默认 1
                            width: 30,
```

```
            shadowBlur: 15,          //阴影面积，即发光效果
            shadowColar: "#000fff"
        }
    },
    radius: "60%",
    //多表盘必配项，center 用来决定表盘的位置
    center: ["50%", "60%"],
    startAngle: 225,
    endAngle: -45,
    clockwise: true,                 //默认就是 true
    axisTick: {                      //小刻度线
        splitNumber: 5,              //默认就是 5
        length: 8,
        lineStyle: {
            color: "#0eee",
            shadowBlur: 5,
            shadowColor: "#0fff",
            width: 1.5,
        }
    },
    splitLine: {                     //大的分割线
        length: 30,
        lineStyle: {
            color: "#000eee",
            opacity: 1,
            width: 2,
        }
    },
    //刻度标签
    axisLabel: {
        distance: 6,                 //与刻度的距离
        color: "blue",
        fontSize: 15,
    },
    //指针
    pointer: {
        length: "70%",
        width: 10
    },
```

```
            itemStyle: {                          //指针样式
                color: "auto",                    //取数值所在区间颜色
                borderWidth: 3,                   //边线宽
                borderColor: "#FFFFCC",  //颜色
                shadowBlur: 10,                   //发光效果
                shadowColor: "orange"
            },
            emphasis: {},
            detail: {                             //下方的数值
                offsetCenter: [0, "60%"],
                color: "black",
                fontSize: 30,
                formatter: "{value}km/h"
            }
        },
        {  //左边仪表盘
            //必配项
            type: "gauge",
            //必配项
            data: [{ name: "转速", value: 1500 }],
            title: {
                offsetCenter: [0, "-40%"]
            },
            //必配项，表示位置
            center: ["30%", "55%"],
            radius: "40%",
            max: 5000,
            splitNumber: 5,
            startAngle: 295,
            endAngle: 55,
            axisLine: {                           //弧线配置
                lineStyle: {
                    shadowBlur: 10,
                    color: [[0.2, "orange"],
                    [0.7, "#000099"],
                    [1.0, "red"]],
                    width: 13
                }
            }
```

```
        },
        { //右边上仪表盘
            //必配项
            type: "gauge",
            //必配项
            data: [{ name: "燃油表", value: 10 }],
            title: {
                //表盘内标题位置
                offsetCenter: [0, "-40%"]
            },
            //必配项，表示位置
            center: ["70%", "55%"],
            startAngle: 125,
            endAngle: 5,
            radius: "40%",
            max: 10,
            splitNumber: 2,
            detail: {
                show: false
            },
            axisLabel: {
                formatter: function (v) {
                    switch (v) {
                        case 0: return "F";
                        case 5: return "M";
                        case 10: return "E";
                    }
                },
                distance: 5
            },
            axisLine: {
                lineStyle: {
                    color: [[0.3, "green"], [1, "#CC0000"]],
                    width: 13,
                    shadowBlur: 10
                }
            },
        },
        { //右边下仪表盘
```

```
                //必配项
                type: "gauge",
                //必配项
                data: [{ name: "水温表", value: 1 }],
                title: {
                    //表盘内标题位置
                    offsetCenter: [0, "40%"]
                },
                //必配项，表示位置
                center: ["70%", "55%"],
                startAngle: 355,
                endAngle: 245,
                radius: "40%",
                max: 10,
                splitNumber: 2,
                detail: {
                    show: false
                },
                axisLabel: {
                    formatter: function (v) {
                        switch (v) {
                            case 0: return "H";
                            case 5: return "M";
                            case 10: return "C";
                        }
                    },
                    distance: 5
                },
                axisLine: {
                    lineStyle: {
                        color: [[0.5, "#000099"], [1, "orange"]],
                        width: 13,
                        shadowBlur: 10
                    }
                },
            }]
        };
        setInterval(function () {   //间隔 1.5 s 更新一次数据
            //改变显示的值，结果保留两位小数
```

```
            var v = ((Math.random() * 200).toFixed(2)).valueOf();
            option.series[0].data[0].value = v;
            var v1 = (v / (option.series[0].max * 0.55)) * option.series[1].max;
            // Math.round()取整函数，四舍五入
            option.series[1].data[0].value = Math.round(v1);
            var v2 = (option.series[2].data[0].value - 1);
            if (v2 < 0)
                v2 = option.series[2].max;
            option.series[2].data[0].value = v2;
            option.series[3].data[0].value = Math.random() * 10;
            console.log(v + ":" + v1);
            /* 第 2 个参数是否不立即更新图表，默认为 false，即同步立即更新。如果为
            true，则会在下一个 animation frame 中才更新图表 */
            mychart.setOption(option, true);
        }, 1500);
    </script>
</body>
</html>
```

以上代码实现的汽车仪表盘面板包含 4 个子仪表盘：中间仪表盘显示车速，范围为 0～200，默认值为 50；仪表盘有 10 个刻度线，分绿色、黄色和红色 3 个颜色区间。左边仪表盘显示转速，范围为 0～5000，默认值为 1500；仪表盘有 5 个刻度线，分橙色、蓝色和黄色 3 个颜色区间。右上方仪表盘显示燃油表，范围为 0～10，默认值为 10；仪表盘有两个刻度线，分绿色和红色两个颜色区间。右下方仪表盘显示水温表，范围为 0～10，默认值为 1；仪表盘有两个刻度线，分蓝色和黄色两个颜色区间。

使用浏览器运行该文件，结果如图 3-3 所示。

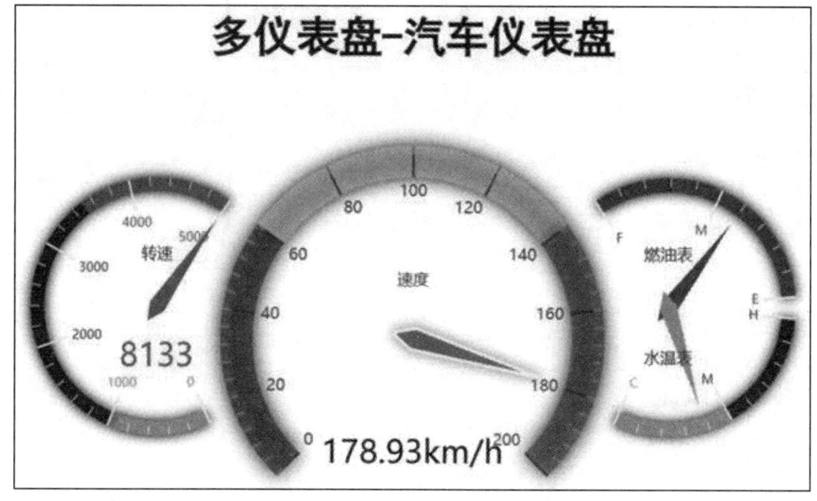

图 3-3　多仪表盘

3.2 绘制漏斗图或金字塔

3.2.1 绘制单个漏斗图

ECharts 中的漏斗图和金字塔图是两种常用的图表类型，它们通常用于展示数据在不同阶段的分布和变化情况。漏斗图是一种倒三角形的条形图，适用于业务流程中各个环节的数据量比较，特别是业务流程规范、周期较长且环节较多的场景，如图 3-4 所示。通过漏斗图，用户可以直观地发现业务流程中问题所在，如在销售流程中分析哪些环节出现客户流失，从而做出相应的策略调整。金字塔图是一种正三角形的条形图，常用于展示数据的层级结构或比重关系。

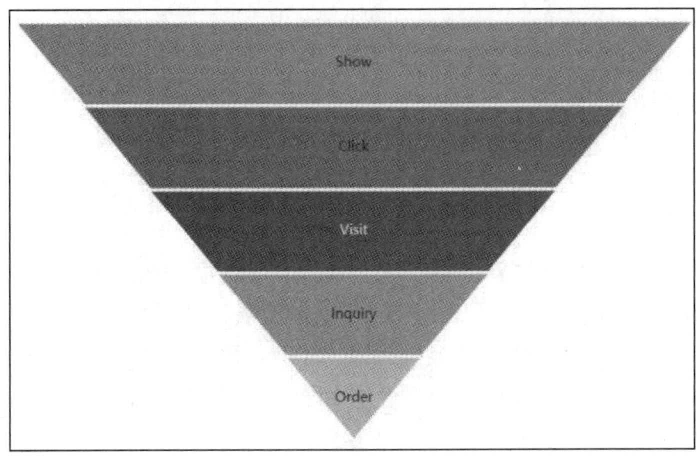

图 3-4 漏斗图效果

使用 ECharts 绘制漏斗图，不需要配置坐标轴，type 属性为 funnel，sort 属性为 descending，数据有 name 和 value 属性。如果要绘制金字塔图，只需要将 sort 属性值改为 ascending。

例如，图 3-4 所示漏斗图的配置项如下：

```
option = {
  title: {
    text: "Funnel",
    left: "center"
  },
  series: [
    {
      name: "Funnel",
      type: "funnel",
      left: "10%",
      top: 60,
```

```
        bottom: 60,
        width: "80%",
        min: 0,
        max: 100,
        minSize: "0%",
        maxSize: "100%",
        sort: "descending",
        gap: 2,
        label: {
          show: true,
          position: "inside"
        },
        itemStyle: {
          borderColor: "#fff",
          borderWidth: 1
        },
        data: [
          { value: 60, name: "Visit" },
          { value: 40, name: "Inquiry" },
          { value: 20, name: "Order" },
          { value: 80, name: "Click" },
          { value: 100, name: "Show" }
        ]
      }
    ]
};
```

上例中还使用了漏斗图的其他常见属性：

(1) min、max：定义了数据的最小值和最大值，这里分别是 0 和 100。

(2) minSize、maxSize：定义了漏斗图每个部分的最小和最大宽度比例。

(3) gap：漏斗图中每个块之间的间隔。

【案例 3-3】 文件 3-2-1.html 实现了一个销售流程转化率的漏斗图，对漏斗图的类型、排序方式、数据等进行基本的配置。

文件 3-2-1.html

```
<!DOCTYPE html>
<html>
<head>
<meta charset="UTF-8">
<title>漏斗图</title>
<script src="../js/echarts.js"></script>
```

```
</head>
<body>
    <div id="main" style="width: 600px; height: 400px"></div>
    <script type="text/javascript">
      var mychart = echarts.init(document.getElementById("main"));
      var option = {
        backgroundColor: "rgba(237, 237, 237, 0.3)",
        title: {
          text: "漏斗图",
          x: "center",
          y:10,
          textStyle:{
            fontSize:30,
            color: "#27408B"
          }
        },
        series: [
          {
            name: "销量",
            //必填项
            type: "funnel",
            // "descending"漏斗, "ascending"金字塔,
            sort: "descending" /* "ascending" ,*/,
            top:"20%",
            //必填项
            data: [
            {
                value: 150,
                name: "浏览商品",
                itemStyle: {
                  normal: {
                    color: "#968FD8",      // 设置区块的颜色
                  },
                },
              },
              {
                value: 120,
                name: "加入购物车",
                itemStyle: {
```

```
            normal: {
              color: "#7986CA",     // 设置区块的颜色
            },
          },
        },
        {
          value: 90,
          name: "提交订单",
          itemStyle: {
            normal: {
              color: "#7FACD6",     // 设置区块的颜色
            },
          },
        },
        {
          value: 60,
          name: "选择支付方式",
          itemStyle: {
            normal: {
              color: "#BFB8DA",     // 设置区块的颜色
            },
          },
        },
        {
          value: 20,
          name: "支付完成",
          itemStyle: {
            normal: {
              color: "#E8B7D4",     // 设置区块的颜色
            },
          },
        }
      ],
      minSize: "5%",                //最小宽度，去除尖角
      maxSize: "100%",
      label: {
        position: /* "inside", */ "left",
      },
      //当标签的位置属性为 "left" 或者 "right" 时才起作用
```

```
            labelLine: {
              length: 40,
              lineStyle: {
                width: 2,
                type: "dashed",            // dashed 由线段组成的虚线
              },
            },
            //设置阴影
            itemStyle: {
              shadowBlur: 10,
              shadowOffsetX: 0,
              shadowOffsetY: 10,
              shadowColor: "rgba(74, 74, 74,0.4)",
            },
            //hover 时的状态，鼠标选中时的状态设置
            emphasis: {
              label: {
                fontFamily: "黑体",
                color: "green",
                fontSize: 28,
              },
            },
          },
        ],
        tooltip: {
          trigger: "item",
          //{a}(系列名称)，{b}(数据项名称)，{c}(数值)，{d}(百分比)
          formatter: "{a}<br/>{b}:{d}%",
        },
        legend: {
          show: true,
          top: 50,
        },
      };
      // 设置"加入购物车"的数据，让其可以每 1.5 s 更新一次
      setInterval (function(){
        option.series[0].data[1].value = Math.round(Math.random() * 50);
        mychart.setOption(option);
      }, 1500);
```

```
    </script>
  </body>
</html>
```

该漏斗图包含 5 个步骤：浏览商品、加入购物车、提交订单、选择支付方式和支付完成，每个步骤的数据值表示该步骤的转化率。通过 setInterval 函数每 1.5 s 更新一次加入购物车的数据值，以模拟实时数据的变化。

使用浏览器运行该文件，结果如图 3-5 所示。

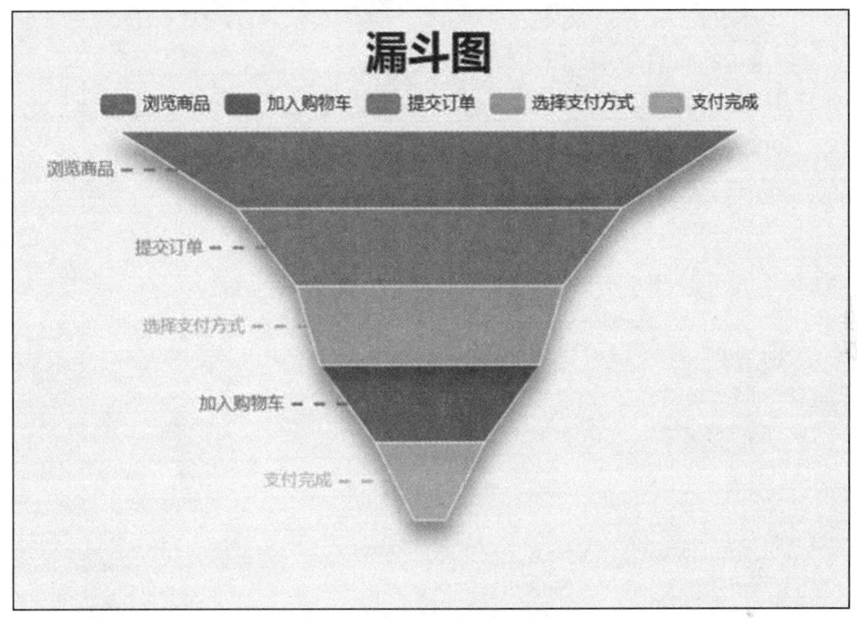

图 3-5　漏斗图案例运行效果

3.2.2　绘制多个图形

单一漏斗图或金字塔图适合展示单个流程或层级结构的细节，而在多个流程或层级结构之间进行比较分析就需要多个图表。绘制多个漏斗图或者金字塔图形，需要配置每个图形的位置属性。

【案例 3-4】 文件 3-2-2.html 实现包含漏斗图和金字塔图两种图形，对两个图形的排序方式、显示位置、数据等进行了基本配置。

文件 3-2-2.html

```
<!DOCTYPE html>
<html>
<head>
<meta charset="UTF-8">
<title>金字塔漏斗图</title>
<script src="../js/echarts.js"></script>
</head>
  <body>
```

```
<div id="main" style="width: 600px; height: 500px"></div>
<script type="text/javascript">
  var mychart = echarts.init(document.getElementById("main"));
  var option = {
    backgroundColor: "rgba(200,200,200,0.1)",
    tooltip: {
      trigger: "item",
      //{a}(系列名称)，{b}(数据项名称)，{c}(数值), {d}(百分比)
      formatter: "{a}<br/>{b}:{d}%",
    },
    title: {
      text: "金字塔漏斗图",
      x: "center",
      y:13
    },
    legend: {
      show: true,
      top: "10%",
    },
    label: {
      position: /* "inside", */ "left",
    },
    labelLine: {
      length: 40,
      lineStyle: {
        width: 2,
        type: "dashed",
      },
    },
    itemStyle: {  //每一条的边框及阴影
      shadowBlur: 10,
      shadowOffsetX: 0,
      shadowOffsetY: 10,
      shadowColor: "rgba(74, 74, 74,0.4)",
    },
    emphasis: {  //hover 时的状态
      label: {
        fontFamily: "黑体",
        color: "green",
```

```
        fontSize: 28,
      },
    },
    series: [
      {
        name: "情况 1",
        type: "funnel",              //必填项
        // "descending"漏斗，"ascending"金字塔
        sort:   "ascending",
        top: "15%",                  //必填项
        height: "40%",               //必填项
        width:"78%",
        x: 10,
        data: [                      //必填项
        {
            value: 150,
            name: "浏览商品",
            itemStyle: {
              normal: {
                color: "#968FD8", //设置每个扇形区块的颜色
              },
            },
          },
          {
            value: 110,
            name: "加入购物车",
            itemStyle: {
              normal: {
                color: "#7986CA", //设置每个扇形区块的颜色
              },
            },
          },
          {
            value: 70,
            name: "提交订单",
            itemStyle: {
              normal: {
                color: "#7FACD6", //设置每个扇形区块的颜色
              },
```

```
                },
            },
            {
                value: 40,
                name: "选择支付方式",
                itemStyle: {
                    normal: {
                        color: "#BFB8DA",        //设置每个扇形区块的颜色
                    },
                },
            },
            {
                value: 20,
                name: "支付完成",
                itemStyle: {
                    normal: {
                        color: "#E8B7D4", //设置每个扇形区块的颜色
                    },
                },
            }
        ],
    },
    {
        name: "情况 2",
        type: "funnel",                    //必填项
        // "descending"漏斗，"ascending"金字塔
        sort: "descending" ,
        top: "55%",                        //必填项
        height: "40%",                     //必填项
        width: "78%",
        x: 10,
        data: [{                           //必填项
                value: 150,
                name: "浏览商品",
                itemStyle: {
                    normal: {
                        color: "#968FD8", //设置每个扇形区块的颜色
                    }
                }
```

```
    },
    {
      value: 110,
      name: "加入购物车",
      itemStyle: {
        normal: {
          color: "#7986CA", //设置每个扇形区块的颜色
        }
      }
    },
    {
      value: 70,
      name: "提交订单",
      itemStyle: {
        normal: {
          color: "#7FACD6", //设置每个扇形区块的颜色
        }
      }
    },
    {
      value: 40,
      name: "选择支付方式",
      itemStyle: {
        normal: {
          color: "#BFB8DA", //设置每个扇形区块的颜色
        }
      }
    },
    {
      value: 20,
      name: "支付完成",
      itemStyle: {
        normal: {
          color: "#E8B7D4", //设置每个扇形区块的颜色
        }
      }
    }
  ]
```

```
            }
          ]
        };
        setInterval (function(){
          // 每 1.5 s 实时更新一次下三角"加入购物车"数据
          option.series[1].data[1].value = Math.round(Math.random() * 80);
          mychart.setOption(option);
        }, 1500);
      </script>
    </body>
</html>
```

以上代码对名称为“情况 1”的漏斗图进行了如下配置：

sort: "ascending" 表示数据值从小到大排序，形成一个金字塔形状；top: "15%" 表示漏斗图距离容器顶部的距离是容器高度的 15%；height: "40%" 表示漏斗图的高度是容器高度的 40%；width: "78%" 表示漏斗图宽度是容器宽度的 78%；x: 10 表示漏斗图距离容器左侧的偏移量为 10 像素。漏斗图“情况 2”与“情况 1”的区别在于 sort: "descending" 表示数据值从大到小排序，形成一个传统的漏斗形状；top: "55%" 表示漏斗图距离容器顶部的距离是容器高度的 55%，最终运行结果如图 3-6 所示。

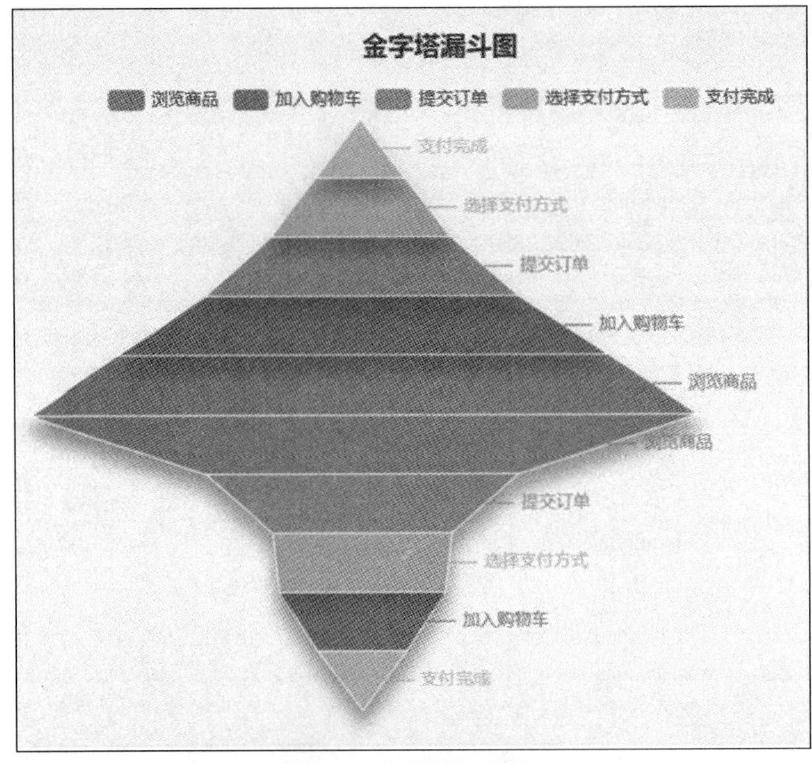

图 3-6 金字塔漏斗图

3.3　ECharts 组件

3.3.1　标题和图例

1. 标题

标题组件 title 如图 3-7 所示，它包含主标题(text)和副标题(subtext)。

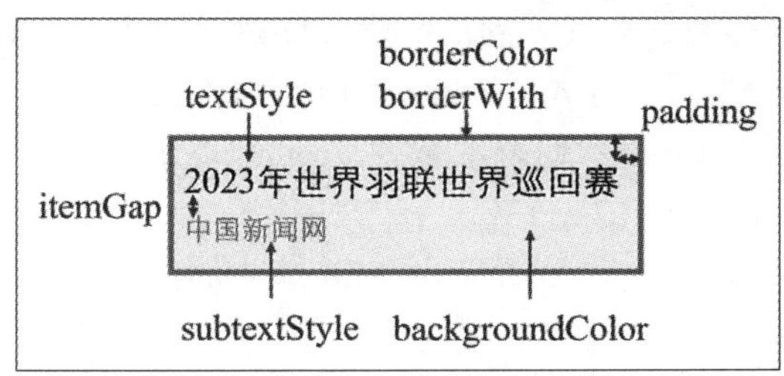

图 3-7　标题示意图

标题和副标题常见属性如下：

- show：是否显示标题组件，boolen 类型。
- text：主标题文本，支持使用\n 换行，string 类型。
- subtext：副标题文本。
- textStyle：对象和 subtextStyle 对象属性。
- color：主标题文字的颜色。
- fontStyle：主标题文字字体风格，可取 normal、italic 和 oblique。
- fontWeight：主标题文字字体粗细，可取 normal、bold、bolder 和 lighter。
- fontSize：主标题文字字体大小。

2. 图例

图例组件如图 3-8 所示，展现了不同系列的标记、颜色和名字，单击图例可以控制对应系列数据显示或不显示。当图例数量过多时，可以滚动图例(垂直/水平)方便查看。

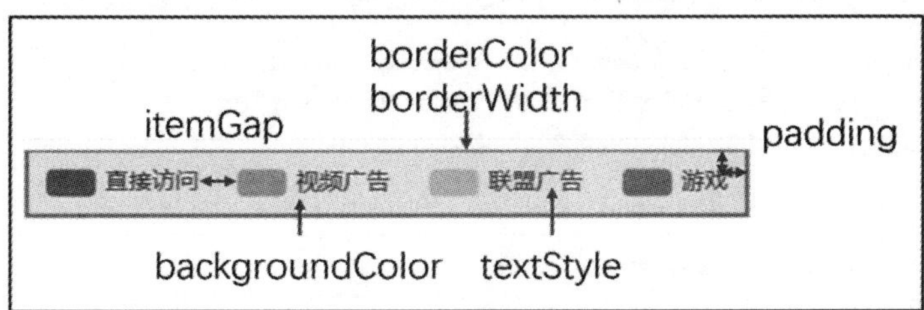

图 3-8　图例示意图

图例属性如下：

• left：图例组件与容器左侧的距离。可使用具体像素，如 left: 20；也可以使用相对于容器宽度的百分比，如 left: "20%"；还可以使用表示水平方向的英文单词，

如center和right。

- top：图例组件与容器上侧的距离。可使用具体像素，如top: 20；也可以使用相对于容器高度的百分比，如top: "20%"；还可以使用表示垂直方向的英文单词，如middle和bottom。
- right：图例组件与容器右侧的距离，与left属性配置一致。
- bottom：图例组件与容器下侧的距离，与top属性配置一致。
- width：图例组件的宽度，默认自适应。
- height：图例组件的高度，默认自适应。
- orient：图例列表的布局朝向。horizontal水平排列，vertical垂直排列。
- align：图例列表对齐方式，一般与orient属性配合使用。默认值auto表示图例列表将自动对齐，left表示列表左对齐，center表示列表居中对齐，right表示列表右对齐。
- padding：图例内边距，单位px，默认各方向内边距为5 px，可使用数组分别设定上右下左边距。
- itemGap：图例每项之间的间隔。横向布局时为水平间隔，纵向布局时为纵向间隔。
- itemWidth：图例标记的图形宽度。
- itemHeight：图例标记的图形高度。

【案例3-5】文件3-3-1.html实现了网站各类别访问量柱状图的标题和图例。

文件3-3-1.html

```
<!DOCTYPE html>
<html>
<head>
<meta charset="UTF-8">
<title>标题和图例</title>
<script src="../js/echarts.js"></script>
</head>
<body>
    <div id="main" style="width: 800px; height: 600px"></div>
    <script type="text/javascript">
        var mychart = echarts.init(document.getElementById("main"));
        var len = 7;
        var xdata = new Array(len);
        for (var i = 0; i < len; i++) {
            xdata[i] = "星期"+(i+1).toString();
        }
        console.log(xdata.join(", "));
        var mydata = new Array(4);
        for (var i = 0; i < 4; i++) {
```

```
        mydata[i] = new Array(len);
        for (var j = 0; j < len; j++) {
            mydata[i][j] = Math.random() * 100;
        }
        console.log(mydata[i].join(", "));
    }
    var option = {
     title : {
            text : "标题组件",
            subtext:"点击标题前往帮助",
            textStyle : {
                color : "blue",
                fontSize : 20,
                fontStyle:"oblique"
            },
            backgroundColor:"yellow",
            borderColor:"#ff0000",
            borderWidth:2,
            shadowBlur:10,
            shadowColor:"blue",
            left:"16%",
            top:5,
            /*标题居中*/
            textAlign : "center",
            link:"https://echarts.apache.org/zh/option.html#title",
            target:"blank",  //打开链接新窗口
       },
       legend:{
          right:"15%" ,
          orient:"horizontal"   /* "vertical" */,
            borderColor:"black",
            borderWidth:2,
            backgroundColor:"rgba(178, 34, 34, 0.2)",
            itemGap:30,
            type:"scroil",  //项目很多时滚动
            padding:10
        },
       xAxis : {
```

```
                data : xdata,
                type:"category",
                splitLine:{
                    show:true
                }
            },
            yAxis :{},
            series : [ {
                    name : "直接访问",
                    type : "bar",
                    data : mydata[0]
                }, {
                    type : "bar",
                    data : mydata[1],
                    name : "视频广告",
                }, {
                    type : "bar",
                    data : mydata[2],
                    name : "联盟广告",
                }, {
                    name:"游戏",
                    type : "bar",
                    data : mydata[3],
                },
            ],
            grid:{
                x2:"15%"  //让整个绘图区域左移
            }
        };
        mychart.setOption(option);
    </script>
</body>
</html>
```

以上配置标题为“标题组件”，副标题为“点击标题前往帮助”，单击副标题可跳转至指定链接。标题文本的样式为蓝色、斜体、字体大小为 20；图例距离右侧为 15%、水平排列方式、边框为黑色、背景为 rgba(178, 34, 34, 0.2)。

使用浏览器运行该文件，最终运行效果如图 3-9 所示。

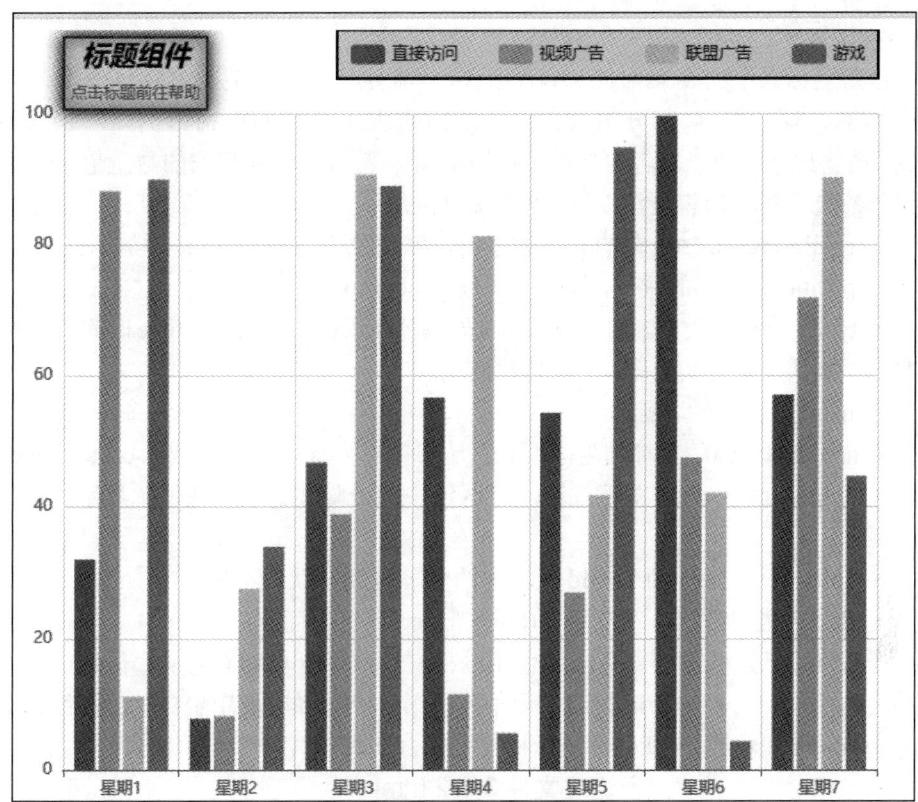

图 3-9　标题和图例案例运行效果

3.3.2　网格和坐标轴

1. 网格

网格即 grid 组件，用于控制图表布局和位置，可将图表分割成多个区域，开发者能够在每个区域内放置不同的图表系列，单个区域内最多可以放置上下两个 x 轴，左右两个 y 轴。其常见属性如下：

- show：是否显示直角坐标系网格。
- left：grid 组件与容器左侧的距离。值可以是具体像素值或相对于容器高宽的百分比，也可以是 center 和 right 等描述。
- top：grid 组件与容器上侧的距离。值可以是具体像素值或相对于容器高宽的百分比，也可以是 middle 和 bottom 等描述。
- width：grid 组件的宽度。
- height：grid 组件的高度。
- backgroundColor：背景色，默认透明。可以使用 RGB 或 RGBA 表示，如“rgb(128, 128, 128)”或“rgba(128, 128, 128, 0.5)”，也可以使用十六进制格式，如“#ccc”。
- borderColor：网格的边框颜色。
- borderWidth：网格的边框线宽。

2. 坐标轴

直角坐标系中的 x 轴对应 xAxis 组件，y 轴对应 yAxis。它们都由轴线、刻度、刻度标签、轴标题 4 部分组成。x 轴常用来表示数据的类别，例如“销售日期”“销售地点”“产品名称”等；y 轴常用来表示某一类数据的数量值，例如“销售数量”和“销售金额”等。常见属性如下：

- gridIndex：x 轴所在的 grid 的索引，默认位于第一个 grid。
- position：x 轴的位置，可为 top 或者 bottom。
- type：坐标轴类型。通常有 value 数值轴、category 类目轴、time 时间轴、log 对数轴，默认 x 轴为类目轴，y 轴为数值轴。
- name：坐标轴名称。
- nameLocation：坐标轴名称显示位置。通常为 start、middle 或者 center、end。
- splitLine：坐标轴刻度线在 grid 区域中的分隔线。默认数值轴显示，类目轴不显示。
- splitArea：坐标轴在 grid 区域中的分隔区域，默认不显示。
- data：类目数据，在类目轴(type: "category")中有效。

【案例 3-6】 文件 3-3-2.html 使用网格实现绘图区域切割为 4 个区域，并绘制不同的图形，第一个 grid 使用两个轴线；第 4 个 grid 中使用坐标轴分割线；对轴线 axisLine 和轴线 axisLabel 进行相关配置，实现轴线与标签的样式。

文件 3-3-2.html

```
<!DOCTYPE html>
<html>
<head>
    <meta charset="UTF-8">
    <title>坐标轴</title>
    <script src="../js/echarts.js"></script>
</head>
<body>
    <div id="main" style="width: 950px; height: 600px"></div>
    <script type="text/javascript">
        var mychart = echarts.init(document.getElementById("main"));
        var len = 7;
        var mydata = new Array(4);
        var tem = [];
        var humidity=[]
        for (var i = 0; i < 4; i++) {
            mydata[i] = new Array(len);
            for (var j = 0; j < len; j++) {
                mydata[i][j] = Math.random() * 100;
            }
```

```
        console.log(mydata[i].join(", "));
    }
    for (var j = 0; j < len; j++) {
        tem[j] = Math.random() * 40;
        humidity[j] = Math.random() * 50;
    }
    var option = {
        grid: [
            { left: "7%", top: "10%", width: "38%", height: "38%" },
            { right: "7%", top: "10%", width: "38%", height: "38%" },
            { left: "7%", bottom: "7%", width: "38%", height: "38%" },
            { right: "7%", bottom: "7%", width: "38%", height: "38%" }
        ],
        xAxis: [
            //第 1 条 x 轴
            {
                // 第二个 grid 中
                gridIndex: 0,
                type: "category",
                position: "top",
                data: ["一", "二", "三", "四", "五", "六", "七"],
            },
            //第 2 条 x 轴
            {
                // 第二个 grid 中
                gridIndex: 0,
                type: "category",
                position: "bottom",
                data: ["一", "二", "三", "四", "五", "六", "七"],
            },
            //第 3 条 x 轴
            {
                // 第二个 grid 中
                gridIndex: 1,
                type: "category",
                data: ["一", "二", "三", "四", "五", "六", "七"],
            },
            //第 4 条 x 轴
            {
```

```
        // 第二个 grid 中
        gridIndex: 2,
        type: "category",
        data: ["一", "二", "三", "四", "五", "六", "七"],
    },
    //第 5 条 x 轴
    {
        // 第一个 grid 中
        gridIndex: 3,
        type: "category",
        position: "bottom",
        //轴线
        axisLine: {
            show: true,
            lineStyle: {
                color: "red",
                type: "solid",
                width: 5
            }
        },
        //刻度
        axisTick: {
            show: true,
            lineStyle: {
                color: "red",
                type: "solid",
                width: 2
            }
        },
        //刻度标签
        axisLabel: {
            show: true,
            interval: "auto",
            rotate: 45,
            margin: 8,
            formatter: "星期{value}"
        },
        //数据，注意中间那个数据字体大
        data: ["1", "2", "3",
```

```
                {
                    value: "4", textStyle: {
                        fontSize: 15
                    }
                }, "5", "6", "7"
            ],
            //分割线
            splitLine: {
                show: true,
                lineStyle: {
                    color: "#483d8b",
                    type: "dashed",
                    width: 2
                }
            },
            //分割区域
            splitArea: {
                show: true,
                areaStyle: {
                    //注意颜色是交叉的
                color: ["rgba(144, 238, 144, 0.3) ", "rgba(135, 200, 250, 0.3) "]
              }
            }
        },
    ],
    yAxis: [
        {
            // 第 1 个 grid 中的 y 轴
            gridIndex: 0,
            type: "value",
            position: "left",
            axisLine: {
                show: true
            }
        },
        {
            // 第 1 个 grid 中的 y 轴
            gridIndex: 0,
            type: "value",
```

```
            position: "right",
            axisLine: {
                linestyle: {
                    color: "red", width: 5
                },
                show: true
            },
            axisLabel: {
                show: true,
                formatter: function (value) {
                    return value + "℃"
                }
            }
        },
        {
            // 第 2 个 grid 中的 y 轴
            gridIndex: 1,
            type: "value",
            axisLine: {
                show: true,
                lineStyle: {
                    color: "blue", width: 5
                }
            },
            axisLabel: {
                formatter: "{value}万次"
            }
        },
        {
            // 第 3 个 grid 中的 y 轴
            gridIndex: 2,
            type: "value",
            axisLine: {
                show: true
            },
            axisLabel: {
                formatter: "{value}万次"
            }
        },
```

```
        {
            // 第 4 个 grid 中的 y 轴
            gridIndex: 3,
            type: "value",
            axisLabel: {
                formatter: "{value}万次"
            }
        },
    ],
    series: [
        {
            name: "气温",
            type: "line",
            smooth: true,
            data: tem,
            //注意这个很关键，因为是两个轴线
            yAxisIndex: 0,
            xAxisIndex: 0,
            lineStyle: {
                color: "red"
            }
        },
        {
            name: "湿度",
            type: "line",
            smooth: true,
            data: humidity,
            //注意这个很关键，因为是两个轴线
            yAxisIndex: 1,
            xAxisIndex: 1,
            lineStyle: {
                color: "green"
            }
        },
        {
            name: "直接访问",
            type: "bar",
            yAxisIndex: 2,
            xAxisIndex: 2,
```

```
                        data: mydata[0]
                    }, {
                        type: "bar",
                        yAxisIndex: 3,
                        xAxisIndex: 3,
                        data: mydata[1],
                        name: "视频广告",
                        stack: "广告"
                    }, {
                        type: "bar",
                        yAxisIndex: 3,
                        xAxisIndex: 3,
                        data: mydata[2],
                        name: "联盟广告",
                        stack: "广告"
                    }, {
                        name: "游戏",
                        type: "bar",
                        yAxisIndex: 4,
                        xAxisIndex: 4,
                        data: mydata[3],
                    },
                ],
                legend: {
                    data: legend_array
                }
            };
            var legend_array = new Array(option.series.length);
            for (var i = 0; i < legend_array.length; i++) {
                legend_array[i] = option.series[i].name;
            }
            mychart.setOption(option);
        </script>
    </body>
    </html>
```

以上代码定义了 4 个网格，意味着将在图表中创建 4 个独立的绘图区域。使用 left、top、right 和 bottom 配置了网格距离容器四周的距离，使用 width 和 height 配置了网格的宽度和高度。

定义了 5 条 x 轴和 5 条 y 轴，gridIndex 指定每条轴都属于一个特定的网格；配置

x 轴的 type 即坐标轴类型都是“category”，表示类目轴；配置 y 轴 type 都为“value”，即数值类型；配置 position 即坐标轴位置；使用 axisLine、axisTick 和 axisLabel 分别配置轴线、轴刻度和轴标签的显示及样式。使用 splitLine 和 splitArea 配置了分割线和分割区域的显示及样式。使用 formatter 配置轴标签的格式化函数，实现自定义显示的文本内容。

使用浏览器运行该文件，最终运行效果如图 3-10 所示。

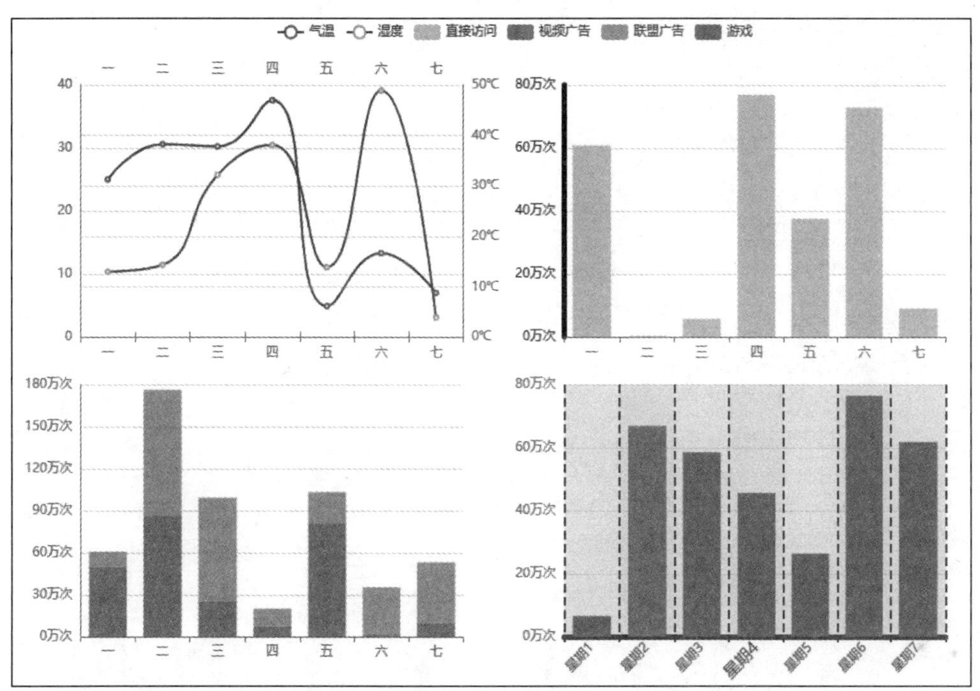

图 3-10　网格和坐标轴案例运行效果

3.3.3　工具箱

ECharts 的 toolbox 为工具箱，该工具箱可为用户提供一系列辅助功能，包括 saveAsImage(保存为图片)、restore(配置项还原)、dataView(数据视图工具)、dataZoom(数据区域缩放)、magicType(动态类型切换)和 brush(选框)共 6 个工具，增强了图表的可操作性和互动性。通过配置这些工具，可以使图表更加符合特定的数据分析需求。具体属性如下：

(1) orient：工具箱 icon 的布局朝向。可选 horizontal 和 vertical。

(2) itemSize：工具箱 icon 的大小。

(3) itemGap：工具箱 icon 每项之间的间隔。横向布局时为水平间隔，纵向布局时为纵向间隔。

(4) showTitle：是否在鼠标 hover 的时候显示每个工具 icon 的标题。

【案例 3-7】 文件 3-3-3.html 实现一个图表工具箱的配置，包含以下功能：保存为图片、配置项还原、数据视图工具、放大/缩小和动态类型切换。

文件 3-3-3.html

```
<!DOCTYPE html>
<html>
<head>
    <meta charset="UTF-8">
    <title>工具箱</title>
    <script src="../js/echarts.js"></script>
</head>
<body>
    <div>
    </div>
    <div id="main" style="width: 800px; height: 500px;"></div>
    <script type="text/javascript">
        var mychart = echarts.init(document.getElementById("main"));
        var len = 25;
        var data_name = ["手机", "空调"];
        var x_label = [];
        var cur = (new Date(2023, 9, 3)).getTime();
        console.log(cur);
        var mydata1 = [];
        var mydata2 = [];
        var i = 0;
        var onday = 24 * 3600 * 1000;
        while (i < len) {
            var now = new Date(cur += onday);
            x_label.push([now.getFullYear(), now.getMonth() + 1, now.getDate()].join("/"))
            mydata1.push(Math.random() * 50 + 10);
            mydata2.push(Math.random() * 10);
            i++;
        }
        console.log(mydata1);
        var option = {
            title: {
                text: "工具箱",
                textStyle: {
                    color: "red",
                    fontSize: 20,
                    fontWeight: "bold"
                }
```

```
    },
    xAxis: {
        type: "category",
        boundaryGap: false,
        data: x_label
    },
    yAxis: {
        type: "value",
    },
    series: [{
        type: "bar",
        smooth: true,
        data: mydata1,
        name: data_name[0],
    }, {
        type: "bar",
        smooth: true,
        data: mydata2,
        name: data_name[1],
    }],
    /*图例*/
    legend: {
        left: 160,
        top: 3
    },
    /*工具箱*/
    toolbox: {
        show: true,
        right: "10%",
        orient: "horizontal",
        backgroundColor: "rgba(0,0,0,0.2) ",
        borderColor: "#ffcccc",
        //鼠标悬停是否显示条目标题
        showTitle: true,
        padding: 5,
        borderWidth: 2,
        feature: {/* 工具配置项*/
            saveAsImage: {/*保存为图片 */
                show: true,
```

```
                        title: "保存我的图片",
                        type: "jpeg"
                    },
                    restore: {/* 配置项还原 */
                        show: true,
                        color: "black"
                    },
                    dataView: {/*数据视图工具*/
                        show: true,
                        title: "数据视图",
                        readOnly: false,
                        lang: ["我的数据视图", "关闭", "刷新"],
                    },
                    dataZoom: {/*放大*/
                        show: true,
                        title: {
                            zoom: "放大",
                            back: "缩小"
                        }
                    },
                    //动态类型切换
                    magicType: {
                        show: true,
                        type: ["line", "bar", "stack"],
                        title: {
                            line: "切换为线性",
                            bar: "切换为柱状",
                            stack: "切换为堆积"
                        }
                    }
                }
            }
        };
        mychart.setOption(option);
    </script>
</body>
</html>
```

该案例代码，配置工具箱包含了保存为图片、还原图表、数据视图、数据缩放和图表类型切换等功能，配置属性包括其显示位置、布局方向、背景和边框样

式。其中位置为图表的右侧，显示为水平方向，背景颜色为半透明的黑色，边框颜色为浅粉色，边框宽度为 2，内边距为 5。

使用浏览器运行该文件，结果如图 3-11 所示。

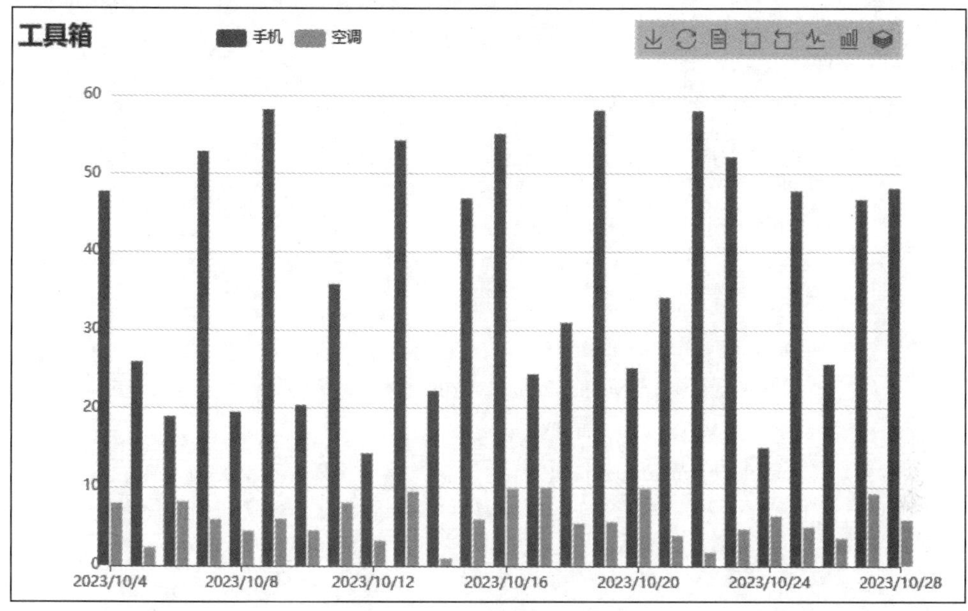

图 3-11　工具箱案例运行结果

3.3.4　提示框

在 ECharts 中，提示框组件 tooltip 是一个非常关键的组件，它主要用于显示图表中数据点的详细信息，通常以弹出框的形式出现。可以设置在全局、坐标系中、系列中、系列的每个数据项中，常用属性如下：

(1) show：是否显示提示框组件，可选为 true 和 false。

(2) trigger：触发类型。item 即数据项图形触发，主要在散点图和饼图等无类目轴的图表中使用；axis 即坐标轴触发，主要在柱状图和折线图等有类目轴的图表中使用；none 即什么都不触发。

【案例 3-8】 文件 3-3-4.html 实现触发类型为坐标轴触发的提示框。

文件 3-3-4.html

```
<!DOCTYPE html>
<html>
<head>
<meta charset="UTF-8">
<title>详情提示框</title>
<script src="../js/echarts.js"></script>
</head>
<body>
    <div id="main" style="width: 800px; height: 600px"></div>
```

```
<script type="text/javascript">
    var mychart = echarts.init(document.getElementById("main"));
    var len = 7;
    var xdata = new Array(len);
    for (var i = 0; i < len; i++) {
        xdata[i] = "星期" + (i + 1).toString();
    }
    console.log(xdata.join(","));
    var mydata = new Array(4);
    for (var i = 0; i < 4; i++) {
        mydata[i] = new Array(len);
        for (var j = 0; j < len; j++) {
            mydata[i][j] = Math.random() * 100;
        }
        console.log(mydata[i].join(","));
    }
    var option = {
        title : {
            text : "详情提示框"
        },
        xAxis : {
            data : xdata
        },
        yAxis : {},
        series : [ {
            name : "直接访问",
            type : "bar",
            data : mydata[0],
            //数据特有的提示框
            tooltip:{
                trigger: "item",
                formatter:"{a}:{b}:{c}"
            }
        }, {
            type : "bar",
            data : mydata[1],
            name : "视频广告",
            stack : "广告"
        }, {
```

```
        type : "bar",
        data : mydata[2],
        name : "联盟广告",
        stack : "广告"
    }, {
        name : "游戏",
        type : "bar",
        data : mydata[3],
    }
    ],
    legend : {},
    //提示框
    tooltip : {
        trigger : "axis",
        //坐标轴指示器
        axisPointer : {
            //类型为十字型，线型为阴影型
            type : /* "cross", *//* "line",  */"shadow",
            //和 type 对应起来后才起作用
            crossStyle : {
                color : "#1e80ff",
                width : 5,
                type : "dashed"
            },
            lineStyle : {
                color : "#1e80ff",
                width : 5,
                type : "solid"
            },
            shadowStyle : {
                color : "rgba(50, 50, 50, 0.1) "
            }
        },
        backgroundColor : "rgba(0, 220, 0, 0.5) ",
        borderColor : "blue",
        borderRadius : 8,
        borderWidth : 2,
        padding:10,
        //显示位置坐标
```

```
                    position:function (p){
                        return [p[0]+10, p[1]-10];
                    },
                    //文本类型
                    textStyle:{
                        color: "blue",
                        fontFamily: "sacns-serif",
                        fontSize:15,
                        fontStyle: "normal",
                        fontWeight: "bold"
                    },
                    //格式化函数形式或者字符串形式
                    formatter:function (params, ticket, callback){
                        console.log(params);
                        console.log(ticket);
                        //数值名称
                        var res="我的详情<br/>"+params[0].name;
                        //显示数值对应关键字和值
                        for(var i=0; i<params.length; i++){
                            res+="<br/>"+params[i].seriesName+":"+params[i].value;
                        }
                        setTimeout(function (){   //回调
                            callback(ticket, res);
                        }, 500);
                        return "loading";          //显示数值之前，要显示的标题
                    }
                }
            };
            mychart.setOption(option);
        </script>
    </body>
</html>
```

以上代码实现了横坐标触发提示框，即鼠标悬停在坐标轴上时显示提示框；提示框的背景颜色为半透明的绿色，边框颜色为蓝色；提示框的内容格式为自定义函数，首先显示“我的详情”和当前数据的名称；在数据加载完成后，通过回调函数更新提示框的内容；在数据加载完成之前，提示框显示“loading”。

使用浏览器运行该文件，运行效果如图 3-12 所示。

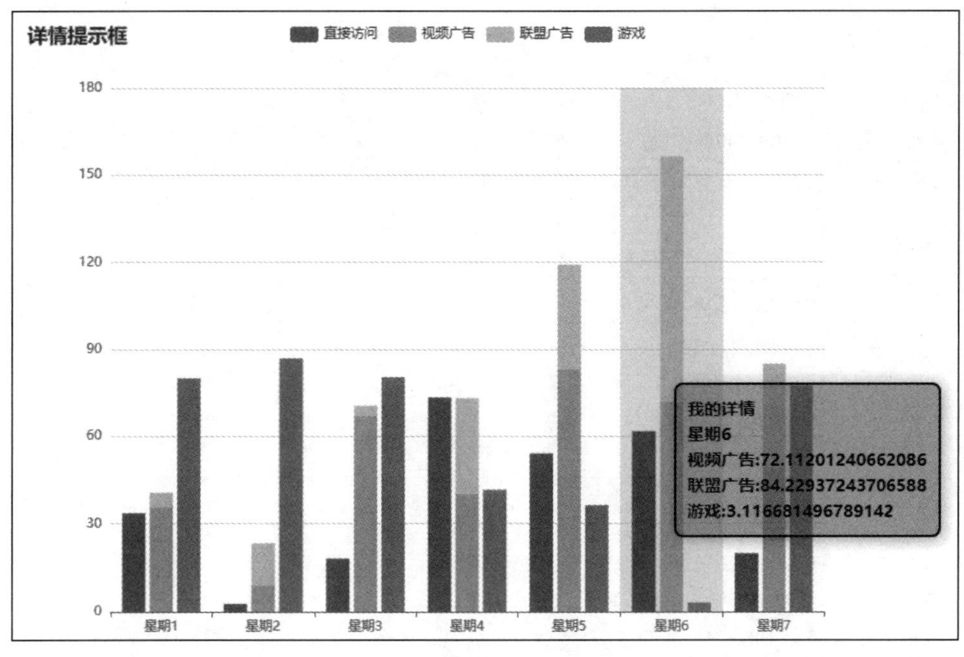

图 3-12 提示框案例运行效果

3.3.5 标记点和标记线

为帮助用户更好地理解和分析数据，需要表示数据的重要点或者参考线，如极值、平均值、特定的数据等，在 ECharts 中可以使用标记点(markPoint)和标记线(markLine)来实现这一功能。

1. 标记点常用属性

(1) symbol 属性即标记点类型，包括 circle、rect、roundRect、triangle、diamond、pin、arrow 和 none。

(2) data 属性即标记点数据，为对象数组。可配置以下两个子属性：

- 位置：可以设置 x 和 y 属性指定相对容器的屏幕坐标，单位像素，支持百分比；也可以设置 coord 属性指定数据在相应坐标系上的坐标位置，单个维度可选 min, max 和 average。
- 类型：用 type 属性标注系列中的最大值和最小值，此时可以使用 valueIndex 或者 valueDim 指定具体维度上的最大值、最小值、平均值。

例如，下面代码定义了 3 个不同的标记点。第 1 个标记点标记数据中的最大值，并显示其名称；第 2 个标记点根据坐标系坐标[1, 20]定位，并显示其名称；第 3 个标记点根据屏幕坐标(x: 100, y: 100)定位，并显示其名称。具体代码如下：

```
markPoint: {
        symbol: "pin",
        data: [
          {
```

```
            name: "最大值",
            type: "max",
            label: {
              formatter: "{b}"
            }
          },
          {
            name: "坐标系坐标",
            coord: [1, 20],
            label: {
              formatter: "{b}"
            }
          },
          {
            name: "屏幕坐标",
            x: 100,
            y: 100,
            label: {
              formatter: "{b}"
            }
          }
        ]
      }
```

具体效果如图 3-13 所示。

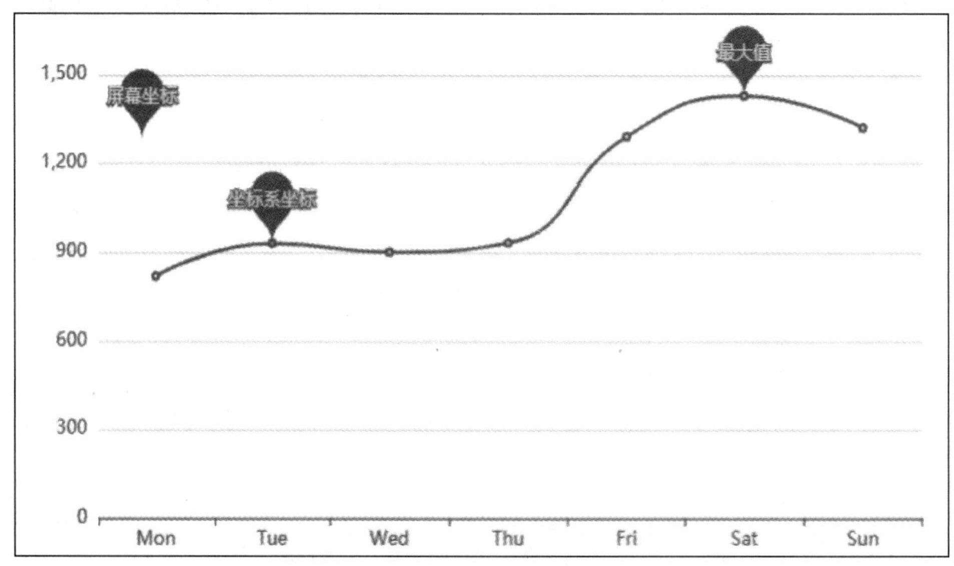

图 3-13 标记点效果

2. 标记线常见属性

• Symbol：标线两端的标记类型，可以使用一个数组分别指定两端，或者单个统一指定。

• symbolSize：标线两端的标记大小，可以使用一个数组分别指定两端，或者单个统一指定。当多个属性同时存在时，优先级顺序是特定于用户当前所使用的图表库和版本，此时应查阅官方文档。

• data：标线的数据数组，数组中包含两个值，分别表示线的起点和终点，每一项是一个对象。

例如，以下代码配置了 4 个标记线，具体代码如下：

```
markLine: {
  data: [
    {
      name: "平均线",
      // 支持 "average", "min", "max"
      type: "average",
      label: { formatter: "{b}{c}" }
    },
    [
      {
        name: "最小值到最大值",
        type: "min",
        label: { formatter: "{b}{c}" }
      },
      { type: "max" }
    ],
    [
      {
        name: "两个坐标之间的标线",
        coord: [1, 120]
      },
      {coord: [1, 310]}
    ],
    [
      {
        name: "两个屏幕坐标之间的标线",
        x: 100,
        y: 100
      },
      { x: 500,y: 200}
```

```
            ]
        ]
    }
```

代码中配置第 1 条标记线为平均值线，标记数据系列的平均值，并显示标签格式为“平均值+数值”；第 2 条标记线是一个范围线，从数据的最小值到最大值；第 3 条标记线是两个坐标系坐标之间的标线，起始于坐标[1, 120]，结束于坐标[1, 310]；第 4 条标记线是两个屏幕坐标之间的标线，起始于屏幕坐标(x: 100, y: 100)，结束于屏幕坐标(x: 500, y: 200)。标记线效果如图 3-14 所示。

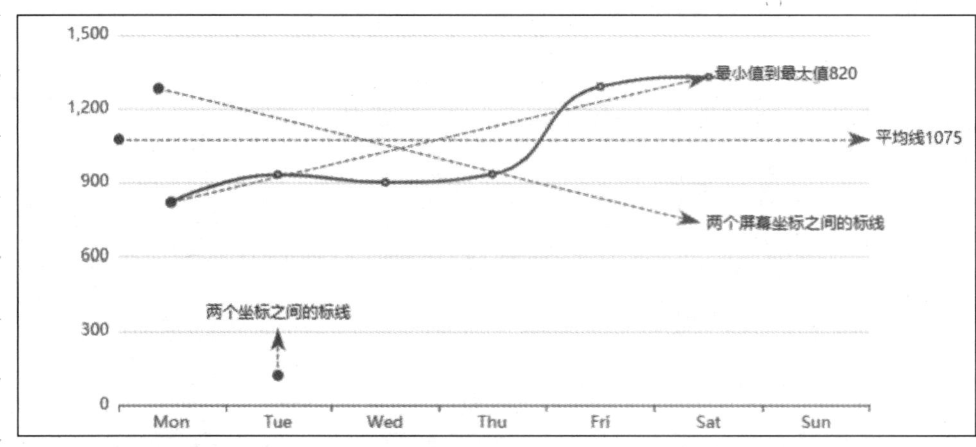

图 3-14 标记线效果

【案例 3-9】 文件 3-3-5.html 实现使用标记点标出最大值和最小值，使用标记线标出平均值。

文件 3-3-5.html

```
<!DOCTYPE html>
<html>
<head>
<meta charset="UTF-8">
<title>标记线及标记点</title>
<script src="../js/echarts.js"></script>
</head>
<body>
    <div id="main" style="width: 600px; height: 400px;"></div>
    <script type="text/javascript">
        var x_string = new Array(12);
        var mydata = [ 3020, 4800, 3600, 6050, 4300, 7200, 4151, 5500, 3020,
                4800, 3600, 6050 ];
        var data_name = "销量";
        for (var i = 0; i < x_string.length; i++) {
```

```
        x_string[i] = i.toString();
    }
    console.log(x_string.join(","));
    var myechart = echarts.init(document.getElementById("main"));
    var option = {
        title : {
            show : true,
            text : "标记点和标记线",
            textStyle : {
                color : "#00ffff"
            }
        },
        legend : {
            show : true
        },
        xAxis : {
            data : x_string
        },
        yAxis : {},
        series : [ {
            name : data_name,
            type : "bar",
            data : mydata,
            markPoint : {  //标记点，注意位置在 serries 里面
                data : [ {
                    type : "max",
                    name : "最大",
                    symbol : "arrow",
                    itemStyle : {
                        color : "red"
                    }
                },
                {
                   type:"min",
                   name:"最小",
                   symbol: "diamond",
                   itemStyle:{
                       color:"yellow"
                   }
                }]
```

```
                },
                markLine:{  //标记线
                    data:[{
                        type:"average",
                        name:"平均值",
                        lineStyle:{
                            color:"darkred",
                            type: "dotted",
                            width:2
                        }
                    }]
                }
            } ]
        };
        myechart.setOption(option);
    </script>
</body>
</html>
```

以上代码中最大值的标记 type 设置为“max”，symbol 设置为“arrow”，表示标记点的符号样式为箭头，itemStyle 中的 color 设置标记点为红色。最小值的标记点 type 设置为“min”，symbol 设置为“diamond”，表示标记点的符号样式为菱形，itemStyle 中的 color 设置标记点为黄色。另外定义了一条用来标记数据平均值的标记线，其 type 设置为“average”，lineStyle 中的 color 设置线的颜色为深红色，type 设置为虚线(“dotted”)。

使用浏览器运行该文件，结果如图 3-15 所示。

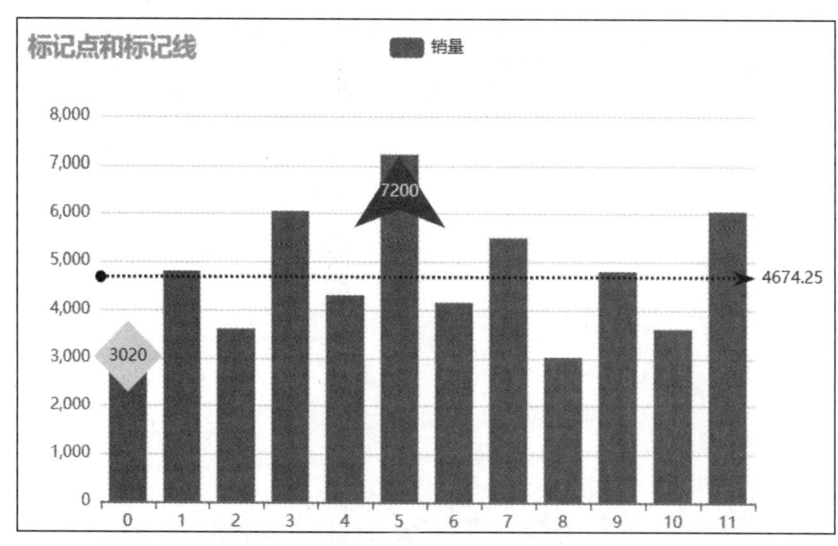

图 3-15　标记点与标记线案例运行结果

3.4　实战演练：电商预警服务网页

1. 需求分析

使用仪表盘展示各商品类别的实时销售数量、销售金额、销售占比等信息；要求根据实际需求定制不规则的刻度，使数据展示更加直观。

使用金字塔图展示各商品类别的销售占比；按照销售金额进行排序，以显示销售额最高的商品类别。

需求分析如下：

序号	需求类别	需 求 描 述
1	仪表盘图	• 确定商品：服装类、数码类、家具类、美妆类、食品类、母婴类，数据均为随机数 • 模拟数据：将销售总额、销售数量、各商品销售数量设置为随机数 • 确定多个图的摆放位置
2	金字塔图	• 确定商品及数据：与环形图数据保持一致 • 设置多个漏斗图：四个图的摆放位置，升序降序设置
3	交互功能	• 实现图表的数据提示，提升用户体验

2. 规划设计

填写如下规划设计表：

<table>
<tr><td colspan="4">WBS 表</td></tr>
<tr><td colspan="4">项目基本情况</td></tr>
<tr><td>项目名称</td><td>电商预警服务网页</td><td>任务编号</td><td></td></tr>
<tr><td>姓　　名</td><td></td><td>班　　级</td><td></td></tr>
<tr><td colspan="4">工作分解</td></tr>
<tr><td>工作任务</td><td colspan="2">包含活动</td><td>备注</td></tr>
<tr><td rowspan="2">1. 环境搭建</td><td colspan="2">1.1　下载 js 包</td><td></td></tr>
<tr><td colspan="2">1.2　拷贝 ECharts 包至本地文件夹</td><td></td></tr>
<tr><td rowspan="5">2. 项目实现</td><td colspan="2">2.1　数据准备</td><td></td></tr>
<tr><td colspan="2">2.2　设置仪表盘的相关配置</td><td></td></tr>
<tr><td colspan="2">2.3　设置漏斗图的相关配置</td><td></td></tr>
<tr><td colspan="2">2.4　设置圆环图的相关配置</td><td></td></tr>
<tr><td colspan="2">2.5　设置提示框</td><td></td></tr>
</table>

3. 搭建环境

(1) 从官网 https://echarts.apache.org/ 下载完整 js 包。

(2) 本地新建文件夹，并拷贝下载的 ECharts 包到 js 文件夹。

4. 项目实现

1) 数据准备

需要初始化数据，用于后续的图表绘制，具体代码如下：

```
var name_data = ["服装类", "食品类", "数码类", "美妆类", "母婴类", "家具类"];
var mydata = new Array();
for (myname in name_data) {
    // Math.trunc() 对小数取整
     mydata.push({
         value: Math.trunc(Math.random() * 100),
         name: name_data[myname],
     });
}
…
//设置间隔 1.5 s 更新一次数据
setInterval(function (){
    …
},1500};
```

以上代码中变量 name_data 存储 6 种商品类型名称，mydata 存储的是字典类型数据，里面包含 6 种商品名称以及每个商品类别对应的销售数据。使用 setInterval() 函数设置销售数量与销售总额每 1.5 s 更新一次。

2) 设置仪表盘的相关配置

在 series 中设置左边和中间仪表盘的配置信息，包括数据、半径、位置、刻度、颜色等，格式参考如下：

```
series: [
    {
    // 设置中间仪表盘相关配置
        type: "gauge",
          …
    },
    {
    // 设置左边仪表盘相关配置
        type: "gauge",
          …
    },
]
```

3) 设置漏斗图的相关配置

在 series 中设置下方漏斗图的配置信息，包括数据、位置、轴线、颜色、排序方式等，格式参考如下：

```
series: [
    {
        // 设置漏斗图相关配置
        type: "funnel",
          …
    },
    {
        type: "gauge",
        // 设置金字塔图相关配置
         …
    },
]
```

4) 设置圆环图的相关配置

在 series 中设置右边圆环图的配置信息，包括数据、半径、位置、颜色等，格式参考如下：

```
series: [
    {
        // 设置圆环图相关配置
        type: " pie",
         …
    },
]
```

5) 设置提示框

在 option 中配置提示框信息，格式参考如下：

```
var option = {
    …
    // 设置提示框
    tooltip:{
        …
    }
}
```

6) 最终 html 文件实现

实现以上 1)～5)个模块代码，然后将其合并。另外添加必备配置(如 js 插件的引入、div 标签的设置、实例的初始化等)，生成文件 saleWarningData.html。

saleWarningData.html

```
<!DOCTYPE html>
<html>
```

```
<head>
    <meta charset="UTF-8">
    <title>多仪表盘</title>
    <!-- 引入在线 ECharts -->
    <script src="https://cdnjs.cloudflare.com/ajax/libs/echarts/4.3.0/echarts.min.js"></script>
</head>
</head>
<body>
    <div id="main" style="width:1200px; height:1000px;"></div>
    <script type="text/javascript">
        var mychart = echarts.init(document.getElementById("main"));
        var name_data = ["服装类", "食品类", "数码类", "美妆类", "母婴类", "家具类"];
        var mydata = new Array();
        for (myname in name_data) {
            // Math.trunc() 对小数取整
            mydata.push({
                value: Math.trunc(Math.random() * 100),
                name: name_data[myname],
            });
        }
        var option = {
            title: {
                text: "销售分析图",
                x: "left",
                y: 15,
                textStyle: {
                    fontFamily: "黑体",
                    fontSize: 30,
                    color: "blue"
                }
            },
            series: [
                //中间仪表盘
                {
                    name: "销售总额",                    //必选
                    //表盘内部的标题，内容同以上的 name
                    title: {
                        offsetCenter: [0, "-20%"],
                        textStyle: {
```

```
                fontFamily: "等线",
                color: "black",
                fontSize: 20,
            }
        },
        type: "gauge",                  //必配项
        data: [{                        //必配项
            name: "销售总额",
            value: 50,
        }],
        min: 0,
        max: 1000,                      //默认值为 100
        splitNumber: 10,                //可以不配置，默认 10
        axisLine: {                     //仪表轮廓线
            show: true,
            lineStyle: {
                //各个区间颜色
                color: [[0.2, "#fc8251"],
                [0.8, "#9A60B4"],
                [1.0, "#3BA272"]],
                opacity: 1,             //透明度默认 1
                width: 15,
                shadowBlur: 10,         //阴影面积，即发光效果
                // shadowColor: "gray"
            }
        },
        radius: "35%",
        //多表盘必配项，center 用来决定表盘的位置
        center: ["41%", "26%"],
        startAngle: 225,
        endAngle: -45,
        clockwise: true,                //默认就是 true
        splitLine: {                    //大的分割线
            length: 8,
            lineStyle: {
                // color: "#000eee",
                opacity: 1,
                width: 2,
                // shadowBlur: 5,
```

```
                    // shadowColor: "gray"
                }
            },
            axisTick: false,
            //刻度标签
            axisLabel: {
                distance: 8,                //与刻度的距离
                color: "blue",
                fontSize: 12,
            },
            //指针
            pointer: {
                length: "60%",
                width: 10
            },
            //指针样式
            itemStyle: {
                color: "auto",              //取数值所在区间颜色
                //描边线宽及颜色
                borderWidth: 1,
                borderColor: "#FFFFFF",
                shadowBlur: 10,             //发光效果
                shadowColor: "gray"
            },
            emphasis: {
            },
            //下方的数值
            detail: {
                offsetCenter: [0, "60%"],
                color: "black",
                fontSize: 20,
                formatter: "{value}万"
            }
        },
        //左边仪表盘
        {
            //必配项
            type: "gauge",
            //必配项
```

```
        data: [{ name: "销售数量", value: 1500 }],
        title: {
            offsetCenter: [0, "-32%"]
        },
        //必配项，表示位置
        center: ["12%", "26%"],
        radius: "25%",
        max: 500,
        splitNumber: 5,
        splitLine: {
            length: 13
        },
        //弧线配置
        axisLine: {
            lineStyle: {
                shadowBlur: 10,
                color: [[0.2, "#5470c6"],
                [0.7, "#ef6567"],
                [1.0, "#f9c956"]],
                width: 10
            }
        },
        //下方的数值
        detail: {
            offsetCenter: [0, "60%"],
            color: "black",
            fontSize: 20,
            formatter: "{value}万"
        }
    },
    //右边饼图
    {
        name: "销量",
        type: "pie",
        label: {
            formatter: "{b}:{d}%",
            fontSize:15
        },
        // "22%":内半径
```

```
            // "25%":外半径
            radius: ["22%", "25%"],
            center: ["79%", "26%"],  // 表盘在水平和垂直的位置
            data: mydata,
            itemStyle: {
                color: function (colors) {
                    var colorList = [
                        "#fc8251",
                        "#5470c6",
                        "#9A60B4",
                        "#ef6567",
                        "#f9c956",
                        "#3BA272",
                    ];
                    return colorList[colors.dataIndex];
                },
                shadowBlur: 10,
                shadowColor: "gray"
            },
        },
        //设置左下漏斗图
        {
            name: "销售额",
            //必填项
            type: "funnel",
            // 设置左下漏斗图位置
            width: "40%",              // 设置左下图的宽度
            height: "15%",             // 设置左下图的高度
            left: "7%",                // 设置左下图距离左边边缘的距离
            top: "60%",                // 设置左下图距离上边边缘的距离
            funnelAlign: "right",      // 设置以漏斗图右侧为轴
            //必填项
            data: mydata,
            gap: 4,
            itemStyle: {
                    color: function (colors) {
                    var colorList = [
                        "#fc8251",
                        "#5470c6",
```

```
                    "#9A60B4",
                    "#ef6567",
                    "#f9c956",
                    "#3BA272",
                ];
                return colorList[colors.dataIndex % colorList.length];
            },
            // 添加阴影
            shadowBlur: 8,
            shadowColor: "rgba(0, 0, 0, 0.25) ",
        },
        label: {
            position: /* "inside", */ "left",
            fontSize:15
        },
        labelLine: {
            length: 60,
            lineStyle: {
                width: 0.5,
                type: "dashed",
            },
        },
    },
    //设置左上金字塔图
    {
        name: "销售额",
        //必填项
        type: "funnel",
        // 设置左上漏斗图位置
        width: "40%",
        height: "15%",
        left: "7%",
        top: "45%",
        sort: "ascending",
        设置以漏斗图右侧为轴
        funnelAlign: "right",
        //必填项
        data: mydata,
        gap: 4,
```

```
                itemStyle: {
                    color: function (colors) {
                    var colorList = [
                        "#fc8251",
                        "#5470c6",
                        "#9A60B4",
                        "#ef6567",
                        "#f9c956",
                        "#3BA272",
                    ];
                    return colorList[colors.dataIndex % colorList.length];
                },
                shadowBlur: 8,
                shadowColor: "rgba(0, 0, 0, 0.25) ",
            },
            label: {
                position: /* "inside", */ "left",
                fontSize:15
            },
            labelLine: {
                length: 60,
                lineStyle: {
                    width: 0.5,
                    type: "dashed",
                },
            },
        },
        {
            name: "销售额",
            //必填项
            type: "funnel",
            //设置右上漏斗图位置
            width: "40%",
            height: "15%",
            left: "50%",
            top: "45%",
            //设置以漏斗图左侧为轴
            funnelAlign: "left",
            //必填项
```

```
            data: mydata,
            gap: 4,
            itemStyle: {
                    color: function (colors) {
                    var colorList = [
                        "#fc8251",
                        "#5470c6",
                        "#9A60B4",
                        "#ef6567",
                        "#f9c956",
                        "#3BA272",
                    ];
                    return colorList[colors.dataIndex % colorList.length];
                },
                shadowBlur: 8,
                shadowColor: "rgba(0, 0, 0, 0.25) ",
            },
            label: {
                position: /* "inside", */ "right",
                fontSize:15
            },
            //当标签的位置属性为“left”或者“right”时才起作用
            labelLine: {
                length: 60,
                lineStyle: {
                    width: 0.5,
                    type: "dashed",
                },
            },
        },
        {
            name: "销售额",
            //必填项
            type: "funnel",
            //设置右下漏斗图位置
            width: "40%",
            height: "15%",
            left: "50%",
            top: "60%",
```

```
                //"ascending"金字塔,
                sort: "ascending" /* "ascending" ,*/,
                //设置以漏斗图左侧为轴
                funnelAlign: "left",
                //必填项
                data: mydata,
                gap: 4,
                itemStyle: {
                        color: function (colors) {
                        var colorList = [
                            "#fc8251",
                            "#5470c6",
                            "#9A60B4",
                            "#ef6567",
                            "#f9c956",
                            "#3BA272",
                        ];
                        return colorList[colors.dataIndex % colorList.length];
                    },
                    shadowBlur: 8,
                    shadowColor: "rgba(0, 0, 0, 0.25) ",
                },
                label: {
                    position: /* "inside", */ "right",
                    fontSize:15
                },
                labelLine: {
                    length: 60,
                    lineStyle: {
                        width: 0.5,
                        type: "dashed",
                    },
                },
            }
        ],
        graphic: {
            type: "text",
            left: "73.4%",
            top: "25%",
```

```
                style: {
                    text: "各商品销售占比",    // 要显示的文本
                    textAlign: "center",
                    fill: "#000",                  // 文字颜色
                    fontSize: 20
                    // 其他样式配置
                }
            },
            tooltip: {
                trigger: "item",
                //{a}(系列名称), {b}(数据项名称), {c}(数值), {d}(百分比)
                formatter: "销售额 ： {c}万",
            },
        };
        //间隔 1.5 s 更新一次数据
        setInterval(function () {
            //改变显示的值，结果为保留两位小数
            var v = ((Math.random() * 500).toFixed(2)).valueOf();
            option.series[0].data[0].value = v;
            var v1 = (v / (option.series[0].max * 0.55)) * option.series[1].max;
            // Math.round()取整函数，四舍五入
            option.series[1].data[0].value = Math.round(v1);
            var v2 = (option.series[2].data[0].value - 1);
            if (v2 < 0)
                v2 = option.series[2].max;
            option.series[2].data[0].value = v2;
            // option.series[3].data[0].value = Math.random() * 10;
            console.log(v + ":" + v1);
            /* 第 2 个参数是否不立即更新图表，默认为 false，即同步立即更新。如果为
            true，则会在下一个 animation frame 中才更新图表 */
            mychart.setOption(option, true);
        }, 1500);
    </script>
</body>
</html>
```

以上代码用仪表盘展示销售总额和销售数量，配置了最大值(max)、分割段数(splitNumber)、轴线样式(axisLine)；使用饼图展示各商品销售占比，配置了内半径(radius)、中心位置(center)；使用漏斗图展示销售额，配置了宽度(width)、高度(height)、位置(left、top)等；使用 setInterval 函数每 1.5 s 更新一次图表数据，并

通过 mychart.setOption 方法应用新的配置选项。

5. 运行结果

使用浏览器运行文件 saleWarningData.html，结果如图 3-16 所示。

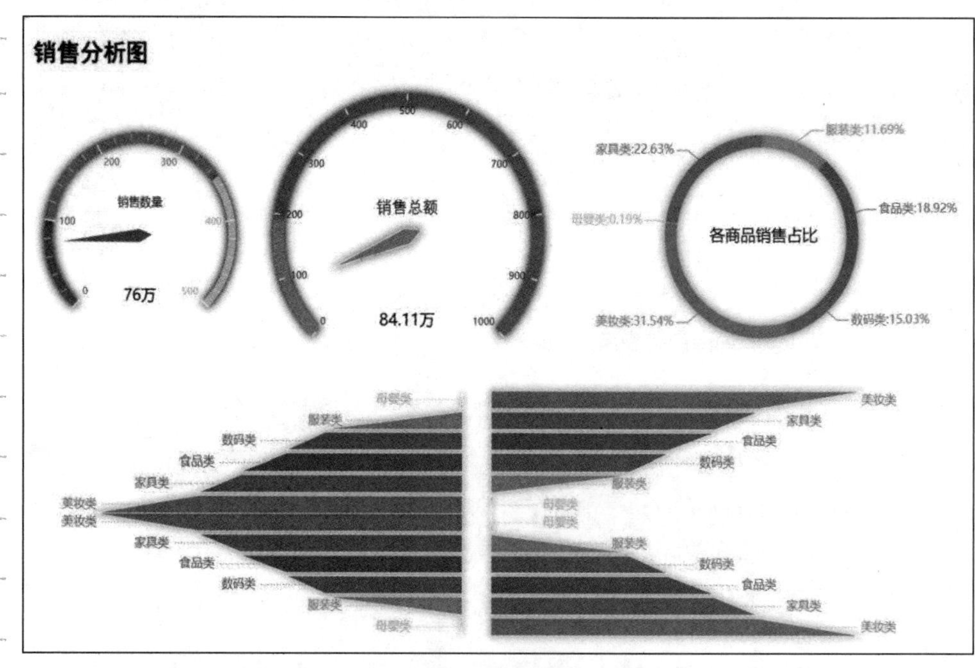

图 3-16　电商预警服务运行结果

通过运行结果可以看出：左边的仪表盘展示商品的销售数量，中间的仪表盘展示商品的销售总额，右边的圆环图展示各商品类型的销售数量占比，下方的漏斗图展示各商品类型的销售额对比，这些图表可以帮助用户更好地分析销售数据的各个方面。

强化练习

习题答案

1. 工具箱 icon 的大小为(　　)。

A. itemGap　　B. showTitle　　C. toolbox.itemSize　　D. time

2. 直角坐标系下有(　　)类型的坐标轴。

A. 类目型　　B. 数值型　　C. 时间型　　D. 其他

3. ECharts 中的工具箱包含(　　)。

A. dataZoom　　B. magicType　　C. saveAsImage　　D. dataView

4. ECharts 中的标记点有(　　)。

A. 最大值　　B. 最小值　　C. 平均值　　D. 任意位置

5. 一张 ECharts 图表一般包含(　　)。

A. 网格区域　　B. x 坐标轴/y 坐标轴

C. 主/副标题　　D. 数据

6. 直角坐标系 grid 中的 x 轴，一般情况下单个 grid 组件最多只能放上下两个 x 轴。(　　)

A. 正确　　B. 错误

7. xAxis.type = "category" 坐标轴类型为数值。(　　)

A. 正确　　B. 错误

8. dataZoom 组件用于区域缩放。(　　)

A. 正确　　B. 错误

9. 工具箱内置有导出图片、数据视图、动态类型切换、数据区域缩放和重置 5 个工具。(　　)

A. 正确　　B. 错误

10. xAxisIndex 为使用的 x 轴的 index。(　　)

A. 正确　　B. 错误

11. 配置图例的选项是(　　)。

A. option　　B. legend　　C. series　　D. tooltip

12. 绘制仪表盘的类型是(　　)。

A. gauge　　B. gaug　　C. gage　　D. gau

13. (　　)用来设置仪表盘的刻度数量。

A. splitNumber　　B. axisLine　　C. splitLine　　D. lineStyle

14. (　　)用来设置仪表盘的大小。

A. offsetCenter　　B. title　　C. center　　D. radius

15. (　　)用来设置仪表盘的位置。

A. radius　　B. center　　C. offsetCenter　　D. width

16. (　　)不是设置颜色的正确方式。

A. color: "rgba(255, 0, 0, 1)"　　B. color: "blue"

C. color: "#0eee"　　D. color: "000eee"

进一步学习建议

学习仪表盘、漏斗图、常用组件之后，可以进一步学习以下内容：

(1) 学习 ECharts 的其他图表类型，如雷达图、盒须图、热力图、路径图、地图等；学习 ECharts 的高级特性，如动态数据交互、动画、主题等。

(2) 关注 ECharts 的更新和发展：ECharts 是一个不断发展和更新的项目，可以关注其官方网站、社交媒体或相关社区，了解其最新的功能、修复及改进。这可以帮助读者及时掌握最新的技术和趋势。

考核评价

<table>
<tr><td colspan="5">考核评价表</td></tr>
<tr><td colspan="2">姓名</td><td></td><td>班　　级</td><td></td></tr>
<tr><td colspan="2">学号</td><td></td><td>考评时间</td><td></td></tr>
<tr><td colspan="2">评价主题及总分</td><td colspan="2">评价内容及分数</td><td>评分</td></tr>
<tr><td rowspan="4">1</td><td rowspan="4">知识考核
(30)</td><td colspan="2">叙述仪表盘的各个组成部分(5分)</td><td></td></tr>
<tr><td colspan="2">论述仪表盘和环形圆饼图的相同与不同之处(5分)</td><td></td></tr>
<tr><td colspan="2">阐述多个漏斗图的位置设置参数(10分)</td><td></td></tr>
<tr><td colspan="2">掌握ECharts常用组件(提示框、工具箱、数据区域缩放等)的功能和使用方法(10分)</td><td></td></tr>
<tr><td rowspan="3">2</td><td rowspan="3">技能考核
(40)</td><td colspan="2">具备业务需求分析、功能设计、编码及调试的综合能力(10分)</td><td></td></tr>
<tr><td colspan="2">按时完成开发任务(20分)</td><td></td></tr>
<tr><td colspan="2">页面呈现的完成度(10分)</td><td></td></tr>
<tr><td rowspan="3">3</td><td rowspan="3">思政考核
(30)</td><td colspan="2">列举仪表盘、漏斗图的其他用途(10分)</td><td rowspan="3"></td></tr>
<tr><td colspan="2">使用仪表盘展示自身数据可视化学习的进度(10分)</td></tr>
<tr><td colspan="2">在解决问题时，能总结经验教训，避免类似问题再次出现(10分)</td></tr>
<tr><td colspan="4">评语：</td><td>汇总：</td></tr>
</table>

第 4 章　Spring 框架

学习目标

目标类型	目 标 描 述
知识目标	• 理解 Spring 框架的基本概况 • 掌握 Spring 框架 Bean 的常见属性 • 理解 Bean 不同实例化方法的区别 • 掌握 Bean 常见注解的作用 • 了解 Spring 框架的作用域和生命周期管理 • 理解 Spring 框架的依赖注入原理和用法 • 理解 AOP 的概念及作用
技能目标	• 能够实现 Bean 的不同作用域配置 • 能够使用 XML、注解方式配置 Spring 框架的 Bean • 能够实现 Spring 框架的 XML 注入，包括构造器注入和设置值注入 • 能够使用 Spring 框架实现 AOP 编程
思政目标	• 培养学习热情和兴趣，提高自主学习能力 • 通过实践掌握 Spring 框架的应用，提升解决实际问题的能力 • 培养团队协作精神，学会与他人分享和交流学习经验 • 树立正确的编程观念，注重代码质量和规范

知识技能储备

4.1　Spring 快速上手

4.1.1　Spring 框架介绍

Spring 是一款开源的 Java 企业级应用程序开发框架，由 Rod Johnson 在 2002 年首次发布，后来由 Pivotal Software 维护和支持。

Spring 框架的主要特点如下：

(1) 控制反转(Inversion of Control，IoC)：传统的 Java EE 开发中代码耦合度较高，需要在代码中创建对象并处理对象之间的依赖关系。而使用了 Spring 框架后，这些将由容器来处理，实现了控制权的反转，使开发者更专注于业务逻辑的实现。

(2) 面向切面编程(Aspect-Oriented Programming，AOP)：在传统的面向对象编程中，通用任务如日志、安全、事务管理等通常会被混杂到业务逻辑中，导致代码混乱而难以维护。Spring 提供了对 AOP 的支持，允许开发者在不修改业务代码的情况下，通过切面对通用任务进行模块化和重用处理。

(3) 集成优秀框架：Spring 框架能够方便地与多种优秀框架集成，如 Struts、Hibernate、Mybatis 等，从而提高软件开发效率和应用程序的性能。

(4) 声明式事务管理：Spring 提供了声明式事务管理，使得开发者能够通过配置文件或注解轻松实现事务控制，无需编写复杂的事务代码。

(5) 降低 Java EE API 使用难度：Spring 提供对 Java EE API 的抽象和封装，简化开发过程，提高开发效率。

(6) 优秀的开源生态：Spring 提供了丰富的与 Spring 框架紧密集成的库、工具和平台，拥有庞大的开发者社区，有助于开发者快速掌握和应用 Spring 框架。

Spring 框架分为核心容器、数据访问/集成层、Web 层和其他模块等，具体如下：

1. 核心容器 CoreContainer

核心容器是 Spring 框架的核心模块，由 Core、Beans、Context、Context-support 和 SpEL 等模块组成。其中 Core 模块提供框架的基本组成部分，包括控制反转(IoC)和依赖注入(Dependency Injection，DI)功能。Spring 将管理对象称为 Bean，Beans 模块提供 Bean 的定义和生命周期的管理，且支持各种 Bean 的作用域。Context 模块是在 Core 和 Beans 模块的基础上建立起来的，提供管理 Bean 的定义、依赖注入、Bean 的生命周期管理、事件发布/订阅等。Context-support 模块使得 Spring 框架能够支持复杂的企业级应用需求，比如国际化支持、资源管理、Web 应用上下文、类型转换、消息源等。SpEL 模块为 Spring 应用程序提供强大的表达式语言，用于如依赖注入、条件注解、事务管理等多种场景。

2. 数据访问/集成层 DataAccess/Intergration

数据访问/集成层包括 JDBC、ORM、JMS 和事务处理模块，用于简化数据库操作、事务管理等。其中 JDBC(Java Database Connectivity)模块可以减少冗长的 JDBC 编码；ORM(Object Relational Mapping)提供对象关系映射 API，Spring 的 ORM 模块支持包括 JPA、MyBatis 和 Hibernate 等多种 ORM 框架；JMS(Java Message Service)模块提供对 Java 消息服务 API 的支持，实现应用程序通过消息通信。Spring 事务处理模块提供对事务管理的全面支持，包括编程式和声明式事务管理。

3. Web 层

Web 层由 Web、Web-MVC、Web-Socket 和 Web-Portlet 等模块组成，其中 Web 模块提供创建 Web 应用的基础设施，包括处理 HTTP 请求和响应的类和方法，多文件上传、Web 应用上下文初始化等；Web-MVC 模块用于构建 Web 应用程序，采用 MVC 设计模式；Web-Socket 模块提供对 WebSocket 协议的支持，允许服务器和客户端之间进行全双工通信；Web-Portlet 模块用于开发和部署 Java Portlet 应用程序，提供 Portlet 环境的 MVC 实现。

4. 其他模块

还有一些其他重要模块，如 AOP、Instrumentation、Test 等模块，其中 AOP 模块通过动态代理实现方法拦截，实现能够在不修改源代码的情况下，增加新的功能；Instrumentation 模块在一定的应用服务器中提供类 Instrumentation 的支持和类加载器的实现；Test 模块支持单元测试和集成测试，支持 Junit 和 TestNG 测试框架。

4.1.2　Spring 入门指南

Spring 入门
实战讲解

为了帮助读者快速掌握 Spring 框架的使用，下面通过一个入门程序来演示 Spring 编程，具体编程步骤如下：

(1) 确保电脑已经安装 JDK1.8 和 IntelliJ IDEA 软件。

(2) 打开 IntelliJ IDEA 软件，选择如图 4-1 所示的新建项目。

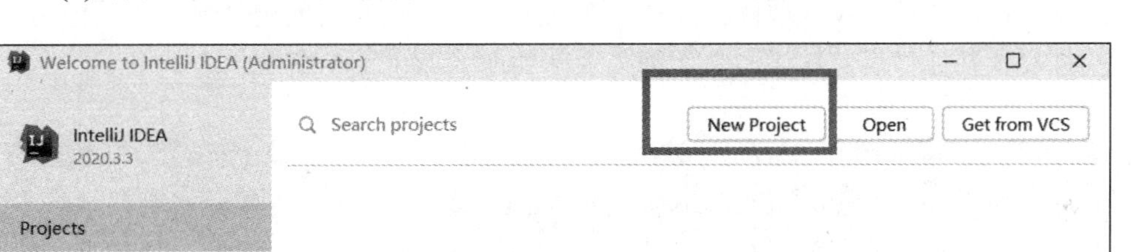

图 4-1　新建项目

(3) 在弹出如图 4-2 所示的对话框内，选择项目类型为“Maven”类型。

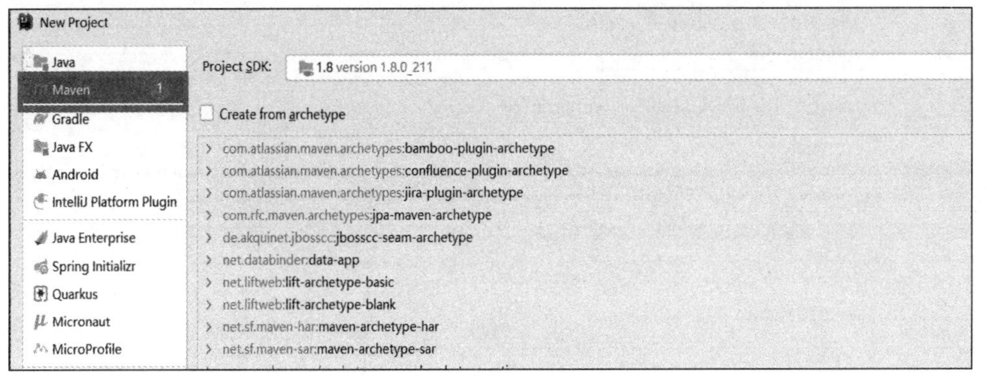

图 4-2　选择项目类型

(4) 在弹出如图 4-3 所示的对话框内，输入项目名称和项目路径后，单击“Finish”按钮完成项目创建。

图 4-3 输入项目名称

(5) 创建完项目后，在自动打开的项目目录中，打开并修改 pom.xml 文件。pom.xml 文件是 Maven 项目的核心配置文件，它定义了项目的构建顺序、依赖关系、插件以及其他信息，这里需要在该文件中加入 Spring 的 6 个依赖 JAR 包。

pom.xml

```
<?xml version="1.0" encoding="UTF-8"?>
<project xmlns="http://maven.apache.org/POM/4.0.0"
         xmlns:xsi="http://www.w3.org/2001/XMLSchema-instance"
         xsi:schemaLocation="http://maven.apache.org/POM/4.0.0
                             http://maven.apache.org/xsd/maven-4.0.0.xsd">
    <modelVersion>4.0.0</modelVersion>
    <groupId>org.example</groupId>
    <artifactId>firstPrj</artifactId>
    <version>1.0-SNAPSHOT</version>
    <properties>
        <maven.compiler.source>8</maven.compiler.source>
        <maven.compiler.target>8</maven.compiler.target>
    </properties>
    <dependencies>
        <dependency>
            <groupId>org.springframework</groupId>
            <artifactId>spring-core</artifactId>
            <version>5.3.24</version>
        </dependency>
```

```
        <dependency>
            <groupId>org.springframework</groupId>
            <artifactId>spring-beans</artifactId>
            <version>5.3.24</version>
        </dependency>
        <dependency>
            <groupId>org.springframework</groupId>
            <artifactId>spring-context</artifactId>
            <version>5.3.24</version>
        </dependency>
        <dependency>
            <groupId>org.springframework</groupId>
            <artifactId>spring-aop</artifactId>
            <version>5.3.24</version>
        </dependency>
        <dependency>
            <groupId>org.springframework</groupId>
            <artifactId>spring-expression</artifactId>
            <version>5.3.24</version>
        </dependency>
        <dependency>
            <groupId>org.springframework</groupId>
            <artifactId>spring-jcl</artifactId>
            <version>5.3.24</version>
        </dependency>
    </dependencies>
</project>
```

对以上文件，新增加的 6 个依赖 JAR 包以及作用，解释如下：

① spring-core：包含 Spring 框架的核心工具类，提供 Spring 核心类如 BeanFactory，以及支持这些类的核心接口和基础功能。

② spring-beans：提供对 Bean 创建、配置、依赖注入(DI)和控制反转(IoC)等相关的功能，Spring 所有应用都要用到该 JAR 包。

③ spring-context：构建在 spring-core 与 spring-beans 之上，提供 Spring 应用程序上下文类 ApplicationContext，还提供对国际化、事件传播、应用层的缓存和 JNDI 查找的支持。

④ spring-aop：这个 JAR 包提供面向切面编程(AOP)的实现，它使用 Java 代理或 CGLIB 来创建代理，实现方法拦截和增强。

⑤ spring-expression：提供对 Spring 表达式语言(SpEL)的支持，用于配置 Spring

应用程序，使其在运行时动态地解析字符串表达式，并将其转换为相应的值或对象属性。

⑥ spring-jcl：提供对处理日志记录的支持，为开发者提供一个统一的日志管理方式，简化日志配置。

(6) 刷新后，项目库文件目录如图 4-4 所示。

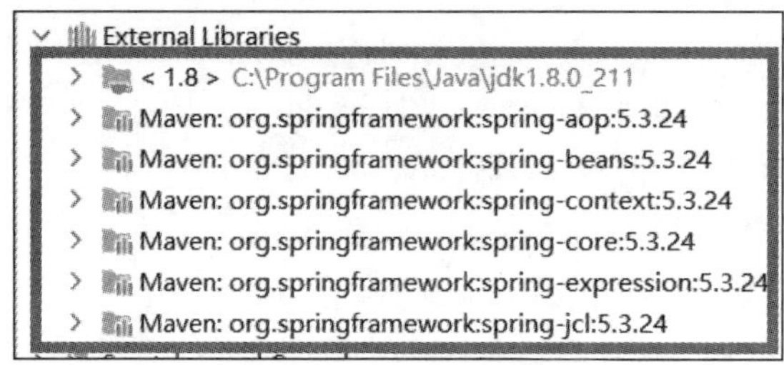

图 4-4　项目库文件目录

(7) 右键单击资源文件目录，新建 Spring 配置文件，如图 4-5 所示。

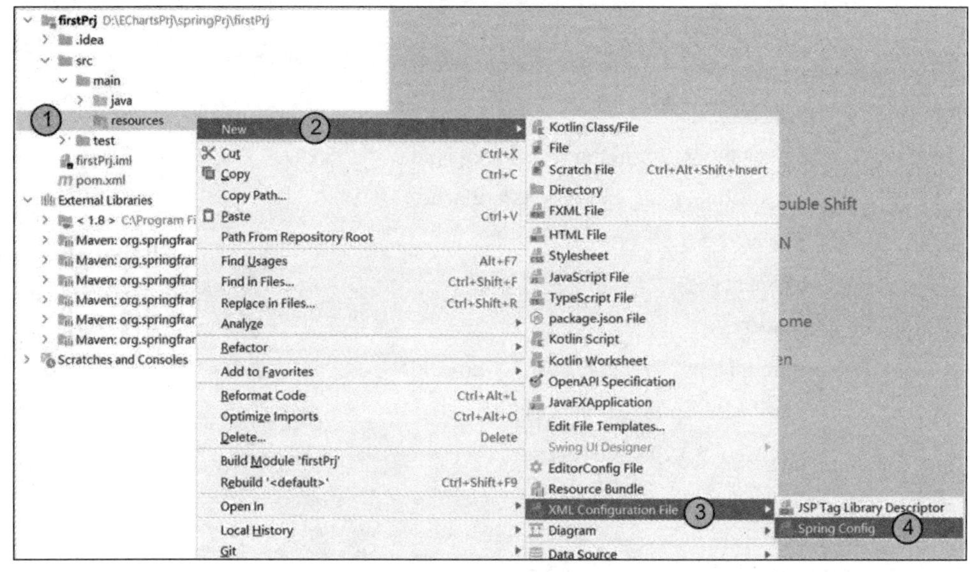

图 4-5　新建 Spring 配置文件

输入文件名 applicationContext，生成 Spring 配置文件，如图 4-6 所示。

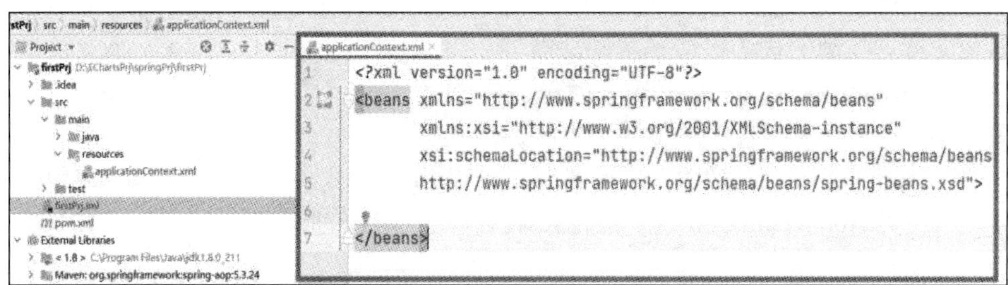

图 4-6　生成的 Spring 配置文件

(8) 在 java 子目录下新建如图 4-7 所示的包 first.user，并创建两个 java 文件，分别为 UserDao.java 和 UserDaoImpl.java。

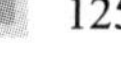

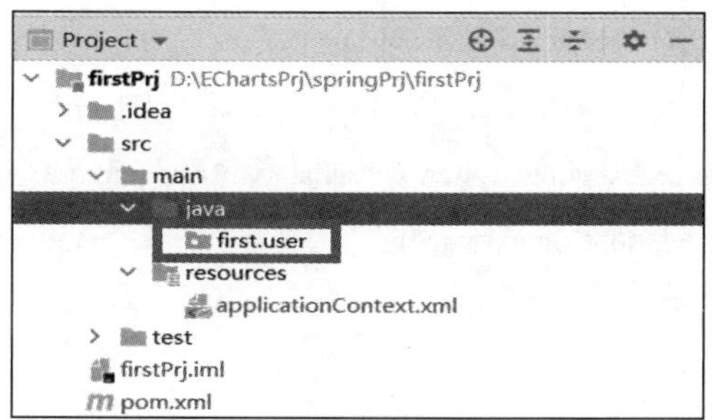

图 4-7　创建包 first.user

UserDao.java

```
package first.user;
public interface UserDao {
    public void say();
}
```

UserDaoImpl.java

```
package first.user;
public class UserDaoImpl implements UserDao {
   @Override
   public void say() {
       System.out.println("helloworld!");
   }
}
```

(9) 修改 Spring 配置文件 applicationContext.xml，添加 id 为 userDao 的 Bean。

在 Spring 框架中，Bean 被定义为在 Spring 容器中管理的对象，Bean 的本质就是 Java 中的类。

applicationContext.xml

```
<?xml version="1.0" encoding="UTF-8"?>
<beans xmlns="http://www.springframework.org/schema/beans"
        xmlns:xsi="http://www.w3.org/2001/XMLSchema-instance"
        xsi:schemaLocation="http://www.springframework.org/schema/beans
        http://www.springframework.org/schema/beans/spring-beans.xsd">
    <bean id="userDao" class="first.user.UserDaoImpl"></bean>
</beans>
```

(10) 在 first.user 包下，新建测试类 Test.java。

Test.java

```
package first.user;
import org.springframework.context.ApplicationContext;
import org.springframework.context.support.ClassPathXmlApplicationContext;
public class Test {
   public static void main(String[] args) {
```

```
//          UserDao userDao=new UserDaoImpl();
//          userDao.say();
        ApplicationContext context=new ClassPathXmlApplicationContext(
                "applicationContext.xml");
        UserDao userDao=    (UserDao) context.getBean("userDao");
        userDao.say();
    }
}
```

在面向对象编程中，为了使用一个类所定义的功能，需要先创建这个类的实例。Spring 容器负责创建和管理对象的生命周期和依赖关系，以上代码中使用 ApplicationContext 从容器中获取对象的实例，即实例化 Bean。

(11) 最终的项目文件结构如图 4-8 所示。

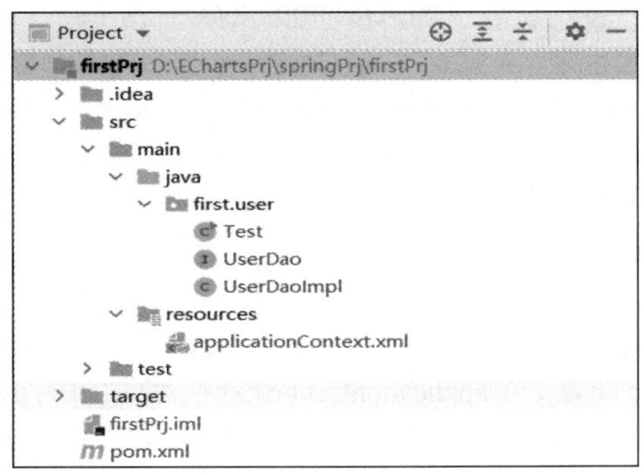

图 4-8 项目文件结构

(12) 运行 Test 类的 main 方法，结果如图 4-9 所示。

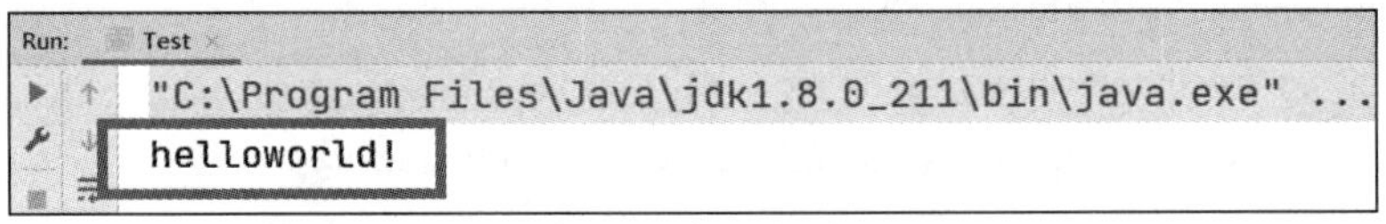

图 4-9 运行结果

以上 Test.java 中屏蔽的代码为传统方式创建的实例对象，而下方为 Spring 方式代码。可以发现使用 Spring 后不需要使用 new 关键字创建对象，而是通过 Spring 的核心容器直接获取对象，代码的控制权发生改变，即实现了控制反转(IoC)。

4.2 Spring 中的 Bean

4.2.1 Bean 的配置和实例化

在 Spring 框架中，Bean 被定义为在 Spring 容器中管理的对象。这些对象不仅包括简单的数据访问对象，也包括复杂的业务逻辑组件，Bean 的本质就是 Java 中的类。

Spring 框架通过其容器来创建、配置和管理 Bean。如果把 Spring 看作是一个工厂，那么 Spring 容器就是生产线，而容器中的 Bean 就是该工厂生产的产品。通过配置 Spring 容器，即告诉它需要创建哪些类型的 Bean 以及如何将这些 Bean 组合在一起工作。

Bean 的常见属性如表 4-1 所示。

表 4-1　Bean 常见属性

属　性	描　　述
id	用于在 Spring 容器中唯一标识一个 Bean 的实例
class	指定用来创建 Bean 的 Java 类，强制性属性
name	名称也是 Bean 标识符，可指定多个名称，名称之间用逗号隔开
scope	作用域，详见 4.2.2 节
constructor-arg	构造器注入依赖关系的参数，详见 4.3.1 节
property	设置值注入依赖关系的参数，详见 4.3.1 节

Bean 的实例化有构造器实例化、静态工厂实例化和实例工厂实例化 3 种方式。

1. 构造器实例化

构造器实例化为最常见的实例化方式，Spring 容器通过反射机制调用类的构造方法来创建对象实例。用户使用 Spring 的构造器实例化方式创建对象，需要基于 XML 配置<bean/>元素，且通过 class 属性指定对象的类型(或类)。

【案例 4-1】 下面通过一个案例介绍构造器实例化的方式，完成多个类的实例化，具体步骤如下：

(1) 参照 4.1.2 节创建 Maven 项目并添加依赖后，创建以下两个类 ThingOne.java 和 ThingTwo.java，并在 say 方法内实现打印输出。

ThingOne.java

```
package demo;
public class ThingOne {
    public void say() {
        System.out.println("this is one");
    }
}
```

ThingTwo.java

```
package demo;
public class ThingTwo {
    public void say() {
        System.out.println("I am two");
    }
}
```

(2) 在配置文件 applicationContext.xml 中，定义两个 Bean，通过 class 属性指定实现类为 ThingTwo 和 ThingOne。

applicationContext.xml

```
<?xml version="1.0" encoding="UTF-8"?>
<beans xmlns="http://www.springframework.org/schema/beans"
    xmlns:xsi="http://www.w3.org/2001/XMLSchema-instance"
    xsi:schemaLocation="http://www.springframework.org/schema/beans
                https://www.springframework.org/schema/beans/spring-beans.xsd">
    <bean id="two" class="demo.ThingTwo"></bean>
    <bean id="one" class="demo.ThingOne"></bean>
</beans>
```

(3) 创建测试文件 Test1.java。

Test1.java

```
package demo;
import org.springframework.context.ApplicationContext;
import org.springframework.context.support.ClassPathXmlApplicationContext;
public class Test1 {
    public static void main(String[] args) {
        ApplicationContext context=new ClassPathXmlApplicationContext(
                "applicationContext.xml");
        ThingOne one= (ThingOne) context.getBean("one");
        one.say();
        ThingTwo two= (ThingTwo) context.getBean("two");
        two.say();
    }
}
```

以上代码先使用 ClassPathXmlApplicationContext 的构造方法实例化核心容器 context，再使用 context 得到 ThingOne 和 ThingTwo 的实例，最后调用这两个实例的 say 方法。

(4) 运行 Test1.java 的 main 方法，结果如图 4-10 所示，从打印结果可以看出，使用构造器实例化方式，Spring 成功创建了 ThingOne 和 ThingTwo 两个类的实例。

```
Test1 ×
"C:\Program Files\Java\jdk1.8.0_211\bin\java.exe" ...
this is one
I am two
```

图 4-10 构造器实例化方式运行结果

2. 静态工厂实例化

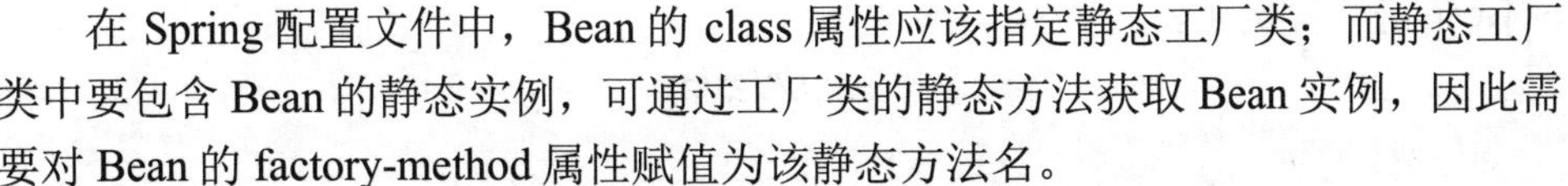

在 Spring 配置文件中，Bean 的 class 属性应该指定静态工厂类；而静态工厂类中要包含 Bean 的静态实例，可通过工厂类的静态方法获取 Bean 实例，因此需要对 Bean 的 factory-method 属性赋值为该静态方法名。

【案例 4-2】 本案例将介绍创建静态工厂类 StaticMethodClass，且使用该类创建 Bean 的实例，具体实现步骤如下：

(1) 参照 4.1.2 节创建 Maven 项目并添加依赖后，创建类 Bean1.java 和静态工厂类 StaticMethodClass.java。

Bean1.java

```
package demo;
public class Bean1 {
    public void say() {
        System.out.println("bean1!");
    }
}
```

StaticMethodClass.java

```
package demo;
public class StaticMethodClass {
    private static Bean1 bean1=new Bean1();
    public static Bean1 createStatic() {
        return bean1;
    }
}
```

在以上代码中，静态工厂类提供了 createStatic 方法，该方法返回 Bean1 的实例。

(2) 修改配置文件 applicationContext.xml，创建要实例化的 Bean，指定静态工厂和工厂方法。

applicationContext.xml

```
<?xml version="1.0" encoding="UTF-8"?>
<beans xmlns="http://www.springframework.org/schema/beans"
      xmlns:xsi="http://www.w3.org/2001/XMLSchema-instance"
      xsi:schemaLocation="http://www.springframework.org/schema/beans
                  https://www.springframework.org/schema/beans/spring-beans.xsd">
    <bean id="bean1"    class="demo.StaticMethodClass"
          factory-method="createStatic"/>
</beans>
```

以上代码中 bean1 的 class 属性定义了静态工厂类的完全限定类名；factory-method 指定用于创建 Bean 实例的工厂方法，这里指定调用类内部的 createStatic 静态方法来创建 Bean。

(3) 创建测试文件 Test2.java，在该类中使用 ApplicationContext 加载配置文件，对 Bean 进行实例化。

Test2.java

```
package demo;
import org.springframework.context.ApplicationContext;
import org.springframework.context.support.ClassPathXmlApplicationContext;
public class Test2 {
    public static void main(String[] args) {
        ApplicationContext context=new ClassPathXmlApplicationContext(
                "applicationContext.xml");
        Bean1 bean1=  (Bean1) context.getBean("bean1");
        bean1.say();
    }
}
```

(4) 运行 Test2.java 的 main 方法，结果如图 4-11 所示，可以看出自定义的静态工厂类 StaticMethodClass 完成了对 Bean1 的实例化。

```
Run:  Test2
"C:\Program Files\Java\jdk1.8.0_211\bin\java.exe" ...
bean1!
```

图 4-11　静态工厂方式运行结果

3. 实例工厂实例化

实例工厂实例化与静态工厂实例化类似，但是会从容器中调用现有 Bean(实例工厂 Bean)的非静态方法来创建新 Bean。

【案例 4-3】 本案例将实现创建实例工厂类 ClassLocator，使用实例工厂创建 Bean2 类的实例，操作步骤如下：

(1) 参照 4.1.2 章节创建 Maven 项目并添加依赖后，创建类 Bean2。

Bean2.java

```
package demo;
public class Bean2{
    public void say() {
        System.out.println("I am Bean2");
    }
}
```

(2) 创建实例工厂类 ClassLocator.java，该类中有一个非静态方法，该方法返回 Bean2 类的实例。

ClassLocator.java

```
package demo;
```

```
public class ClassLocator {
    Bean2 bean2=new Bean2();
    public Bean2 createFactoryClass () {
        return bean2;
    }
}
```

(3) 修改配置文件 applicationContext.xml，先配置实例工厂类的 Bean(id 为 classLocator)，然后创建另外一个 Bean(id 为 bean2)，其 factory-bean 属性配置为实例工厂 Bean 的 id，factory-method 属性指向实例工厂的方法名称。

applicationContext.xml

```
<?xml version="1.0" encoding="UTF-8"?>
<beans xmlns="http://www.springframework.org/schema/beans"
        xmlns:xsi="http://www.w3.org/2001/XMLSchema-instance"
        xsi:schemaLocation="http://www.springframework.org/schema/beans
                        https://www.springframework.org/schema/beans/spring-beans.xsd">
    <bean id="classLocator" class="demo.ClassLocator"></bean>
    <bean id="bean2" class="demo.Bean2"
        factory-bean="classLocator" factory-method="createFactoryClass">
    </bean>
</beans>
```

(4) 创建测试文件 Test.java，在该文件中实现 Bean2 实例的获取和使用。

Test.java

```
package demo;
import org.springframework.context.ApplicationContext;
import org.springframework.context.support.ClassPathXmlApplicationContext;
public class Test3 {
    public static void main(String[] args) {
        ApplicationContext context=new ClassPathXmlApplicationContext(
                "applicationContext.xml");
        Bean2 bean2 = (Bean2) context.getBean("bean2");
        bean2.say();
    }
}
```

(5) 运行 Test.java 的 main 方法，结果如图 4-12 所示，从打印结果可以看出完成了对 Bean2 的实例化。

```
Run:    Test3
"C:\Program Files\Java\jdk1.8.0_211\bin\java.exe" ...
I am Bean2
```

图 4-12　实例工程方式运行结果

4.2.2 Bean 的作用域和生命周期

Bean 定义和配置用于创建一个类的实例，其中 Bean 的作用域(即 scope 属性)决定创建对象实例为一个或者多个。Bean 的生命周期包括 Bean 的定义、Bean 的初始化、Bean 的使用和 Bean 的销毁。由 Spring 容器管理的 Bean，作用域决定 Bean 实例的创建和销毁方式。

(1) singleton：默认作用域配置，表示该 Bean 实例在容器启动时创建，并且以单例方式存在，直到容器关闭时才会被销毁。由于单例方式为默认模式，下面两种方式的定义是等效的：

```
<bean id="accountService" class="com.AccountService"/>
<bean id="accountService" class="com.AccountService" scope="singleton"/>
```

(2) prototype：表示每次从容器中获取 Bean 时，都返回一个新的实例。当容器创建 Bean 实例后，Bean 的销毁由客户端负责，Spring 容器不再负责管理。该作用域配置方法如下：

```
<bean id="accountService" class="com.AccountService" scope="prototype"/>
```

(3) request：表示每次 HTTP 请求开始都会创建一个新的 Bean，在请求结束时销毁 Bean。

(4) session：表示在一个用户会话(HTTP Session)的生命周期内仅有一个 Bean 实例，不同用户会话中存在不同的 Bean 实例，Bean 实例随着用户会话的结束而被销毁。

(5) application：表示在一个 ServletContext 的生命周期内仅有一个 Bean 实例，不同 ServletContext 中存在不同的 Bean 实例，仅适用于 Web 相关的 ApplicationContext 环境。

(6) websocket：表示在 Websocket 的生命周期内仅有一个 Bean 实例，不同 Websocket 中存在不同的 Bean 实例，仅适用于 Web 相关的 ApplicationContext 环境。

【案例 4-4】使用 Spring 创建 singleton 和 prototype 两种作用类型的 Bean，实现从创建到销毁期间观察 Spring 对其生命周期的管理，实现步骤如下：

(1) 参照 4.1.2 节创建 Maven 项目，在工程的 java 目录下创建包 demo，并在该包下创建文件 SingleTonClass.java 和 ProtypeClass.java，以及 Spring 配置文件 beans.xml。

SingleTonClass.java

```
package demo;
public class SingleTonClass {
    public void init() {
        System.out.println("initial bean SingleTonClass");
    }
    public void destroy() {
```

```
            System.out.println("destroy bean SingleTonClass");
        }
    }
```

ProtypeClass.java

```
package demo;
public class ProtypeClass {
    public void init() {
        System.out.println("initial bean ProtypeClass");
    }
    public void destroy() {
        System.out.println("destroy bean ProtypeClass");
    }
}
```

beans.xml

```
<?xml version="1.0" encoding="UTF-8"?>
<beans xmlns="http://www.springframework.org/schema/beans"
    xmlns:xsi="http://www.w3.org/2001/XMLSchema-instance"
    xsi:schemaLocation="http://www.springframework.org/schema/beans
                        https://www.springframework.org/schema/beans/spring-beans.xsd">
    <bean id="single" class="demo.SingleTonClass"
        init-method="init" destroy-method="destroy"></bean>
    <bean id="prototype" class="demo.ProtypeClass"
        scope="prototype" init-method="init" destroy-method="destroy"></bean>
</beans>
```

以上配置文件 beans.xml 中，对于第 1 个 id 为 single 的 Bean，配置 scope 属性为默认值 singleton；第 2 个 id 为 prototype 的 Bean，配置 scope 属性为 prototype；通过 init-method 属性指定一个初始化方法，该方法在 Bean 创建后立即被调用；通过 destroy-method 属性指定一个销毁方法，该方法在 Bean 被销毁前调用。

(2) 在 demo 包下创建测试类 Test.java。

Test.java

```
package demo;
import org.springframework.context.ApplicationContext;
import org.springframework.context.support.AbstractApplicationContext;
import org.springframework.context.support.ClassPathXmlApplicationContext;
public class Test {
    public static void main(String[] args) {
        AbstractApplicationContext context = new ClassPathXmlApplicationContext("beans.xml");
        SingleTonClass s1= context.getBean("single", SingleTonClass.class);
        SingleTonClass s2= context.getBean("single", SingleTonClass.class);
```

```
        System.out.println("s1==s2:"+(s1==s2));
        ProtypeClass p1=context.getBean("prototype", ProtypeClass.class);
        ProtypeClass p2=context.getBean("prototype", ProtypeClass.class);
        System.out.println("p1==p2:"+(p1==p2));
        System.out.println(p1);
        System.out.println(p2);
        //注册一个钩子函数，这个函数会在应用程序关闭时执行清理工作
        context.registerShutdownHook();
    }
}
```

以上代码获取了 id 为 single 的 Bean，根据配置文件可知其 scope 属性为 singleton；获取了 id 为 prototype 的 Bean，根据配置文件可知其 scope 属性为 prototype。

(3) 运行 Test.java 的 main 方法，结果如下：

```
initial bean SingleTonClass
s1==s2:true
initial bean ProtypeClass
initial bean ProtypeClass
p1==p2:false
demo.ProtypeClass@5bda8e08
demo.ProtypeClass@1e800aaa
destroy bean SingleTonClass
```

从以上运行结果可以看出，对于 singleton 作用类型的 Bean，从创建到销毁，Spring 完全管理其整个生命周期；而对于 prototype 作用类型的 Bean，Spring 并没有负责销毁 Bean，而是将其交给客户端。

4.3 依赖注入

Bean 的依赖注入(DI)是定义对象依赖关系的一种过程，也称为 Bean 的装配。Bean 依赖注入的常见方式有 XML 注入方式、注解装配方式和自动装配方式三种。

4.3.1 Bean 的 XML 注入方式

Bean 的 XML 注入方式又分为基于构造器的依赖注入和基于设置值的依赖注入，具体介绍如下。

1. 基于构造器的依赖注入

这种方式通过容器调用带参数的构造器来实现，每个参数代表一个依赖项。例如，对于以下 ThingOne 类，其构造方法中有 ThingTwo 和 ThingThree 类两个类型的参数：

```
package x.y;
public class ThingOne {
    public ThingOne(ThingTwo thingTwo, ThingThree thingThree) {
        // ...
    }
}
```

基于构造器依赖注入的配置方法如下：

```
<beans>
    <bean id="beanOne" class="x.y.ThingOne">
        <constructor-arg ref="beanTwo"/>
        <constructor-arg ref="beanThree"/>
    </bean>
    <bean id="beanTwo" class="x.y.ThingTwo"/>
    <bean id="beanThree" class="x.y.ThingThree"/>
</beans>
```

其中，constructor-arg 代表构造器参数，它默认使用参数的类型匹配。若 ThingTwo 和 ThingThree 类之间没有继承关系，则在 Bean 中定义的构造器参数中不存在歧义；否则需要显式指定构造函数参数的索引(index)或类型(type)，具体代码如下：

```
<beans>
    <bean id="beanOne" class="x.y.ThingOne">
        <constructor-arg  index="0"  ref="beanTwo"/>
        <constructor-arg  index="1" ref="beanThree"/>
    </bean>
    <bean id="beanTwo"  class="x.y.ThingTwo"/>
    <bean id="beanThree"  class="x.y.ThingThree"/>
</beans>
```

多学一招：如果要把一个引用传递给一个对象，需要使用标签的 ref 属性；而如果要直接传递一个值，应该使用 value 属性。

【案例 4-5】有两个实体类 Person 和 Card，Person 实体内有一个 nameString、age 和 card 成员对象，Person 实体依赖于 Card 实体，使用构造器注入完成 Person 类的装配，具体实现步骤如下：

(1) 参照 4.1.2 节完成步骤(1)～(8)后，在 java 目录下创建包 demo4_3_1。

(2) 该包下创建 Person 类和 Card 类，分别为 Person.java 和 Card.java，且 Person 类有一个 Card 类型的成员。

Person.java

```
package demo4_3_1;
public class Person {
    String  nameString;
```

```
    int  age;
    Card card;
    public Person(String nameString, int age, Card card) {
        this.nameString = nameString;
        this.age = age;
        this.card = card;
    }
    @Override
    public String toString() {
        return "Person{" +
                "nameString='" + nameString + '\'' +
                ", age=" + age +",\n"+
                "card=" + card.toString() +
                "}";
    }
}
```

Card.java

```
package demo4_3_1;
public class Card {
    private String cardNumber;
    private String expiryDate;
    public Card(String cardNumber, String expiryDate) {
        this.cardNumber = cardNumber;
        this.expiryDate = expiryDate;
    }
    @Override
    public String toString() {
        return "Card{" +
                "cardNumber='" + cardNumber + '\'' +
                ", expiryDate='" + expiryDate + '\'' +
                '}';
    }
}
```

(3) 创建 Spring 配置文件 beans.xml，实现构造器方式注入。

beans.xml

```
<?xml version="1.0" encoding="UTF-8"?>
<beans xmlns="http://www.springframework.org/schema/beans"
       xmlns:xsi="http://www.w3.org/2001/XMLSchema-instance"
```

```
        xsi:schemaLocation="http://www.springframework.org/schema/beans
                        http://www.springframework.org/schema/beans/spring-beans.xsd">
    <bean id="card" class="demo4_3_1.Card">
        <constructor-arg name="cardNumber" value="1234567"/>
        <constructor-arg name="expiryDate" value="2025-03-01"/>
    </bean>
    <bean id="person" class="demo4_3_1.Person">
        <constructor-arg index="0" value="李明"/>
        <constructor-arg index="1" value="18"/>
        <constructor-arg index="2" ref="card"/>
</bean>
</beans>
```

在以上配置文件中定义了一个 id 为 card 的 Bean，指定了 Card 类构造器的第 1 个参数 cardNumber，并将其值设置为 1234567；Card 类构造器的第 2 个参数 expiryDate，并将其值设置为 2025-03-01。同样，定义了一个 id 为 person 的 Bean，并指定 Person 类构造器的 3 个参数值。

(4) 创建测试类文件 Test.java，在该类中通过 Spring 容器获取 Person 类实例。

Test.java

```
package demo4_3_1;
import org.springframework.context.ApplicationContext;
import org.springframework.context.support.ClassPathXmlApplicationContext;
public class Test {
    public static void main(String arg[]) {
        ApplicationContext context=new ClassPathXmlApplicationContext("beans.xml");
        Person person=context.getBean("person", Person.class);
        System.out.println(person.toString());
    }
}
```

(5) 运行 Test.java 的 main 方法，结果如图 4-13 所示，可以看出使用构造器注入进行 Person 类的装配，完成了 Person 实体类的依赖注入。

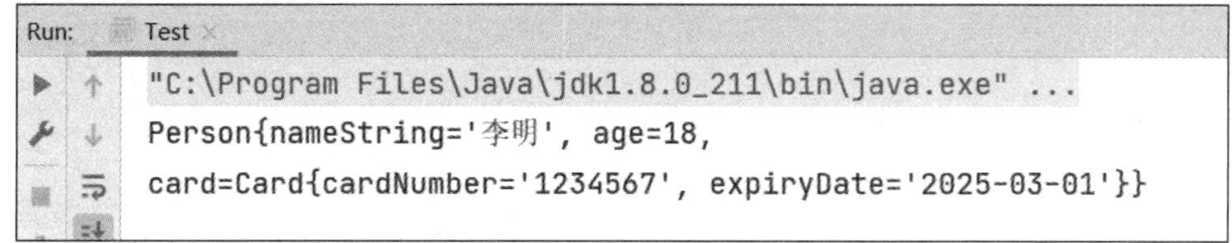
Run:
Test
"C:\Program Files\Java\jdk1.8.0_211\bin\java.exe" ...
Person{nameString='李明', age=18,
card=Card{cardNumber='1234567', expiryDate='2025-03-01'}}

图 4-13　构造器依赖注入案例运行结果

2. 基于设置值的依赖注入

基于设置值的依赖注入是由容器通过调用无参构造函数，或无参数静态工厂

方法实例化 Bean，然后调用 Bean 的 setter 方法来完成的，与基于构造器注入方式的区别在于该方式使用的是<bean>标签的 property 属性，具体代码如下：

```
<bean id="score" class="demo.Score">
        <property name="ch" value="89" />
        <property name="en" value="98" />
</bean>
```

如果想传递多个值，Spring 提供 4 种类型的集合元素：

(1) <list>：表示注入一列值，允许重复。

(2) <set>：表示注入一组值，但不能重复。

(3) <map>：表示注入“名称-值”的集合，其中名称和值可以是任何类型。

(4) <props>：表示注入“名称-值”的集合，其中名称和值都是字符串类型。

【案例 4-6】 本案例中 SetClass 类有 avg、age 等普通基本数据类型的成员，还有成绩类对象、数组、集合等成员，利用 Spring 完成基于设置值的依赖注入，步骤如下：

(1) 参照 4.1.2 节创建 Maven 项目，在 java 目录下创建包 demo4_3_1，该包下创建 ScoreClass 类和 SetClass 类，分别为 ScoreClass.java 和 SetClass.java。注意需要实现对应成员的 setter 方法。

ScoreClass.java

```
package demo4_3_31;
public class ScoreClass {
    int en;
    int ch;
    int math;
    public int getEn() {
        return en;
    }
    public void setEn(int en) {
        this.en = en;
    }
    public int getCh() {
        return ch;
    }
    public void setCh(int ch) {
        this.ch = ch;
    }
    public int getMath() {
        return math;
    }
    public void setMath(int math) {
```

```
            this.math = math;
        }
}
```

SetClass.java

```
package demo4_3_1;
import java.util.Arrays;
import java.util.List;
import java.util.Map;
import java.util.Properties;
import java.util.Set;
public class SetClass {
    float avg;
    int age;
    String nameString;
    ScoreClass scoreClass;
    List<Integer> scorelist;
    String[] books;
    Set gender;
    Map schoolInfoMap;
    Properties familyProperties;
    public int getAge() {
        return age;
    }
    public void setAge(int age) {
        this.age = age;
    }
    public String getNameString() {
        return nameString;
    }
    public void setNameString(String nameString) {
        this.nameString = nameString;
    }
    public ScoreClass getScoreClass() {
        return scoreClass;
    }
    public void setScoreClass(ScoreClass scoreClass) {
        this.scoreClass = scoreClass;
    }
    public List getScorelist() {
```

```
        return scorelist;
    }
    public void setScorelist(List scorelist) {
        this.scorelist = scorelist;
    }
    public String[] getBooks() {
        return books;
    }
    public void setBooks(String[] books) {
        this.books = books;
    }
    public Set getGender() {
        return gender;
    }
    public void setGender(Set gender) {
        this.gender = gender;
    }
    public Map getSchoolInfoMap() {
        return schoolInfoMap;
    }
    public void setSchoolInfoMap(Map schoolInfoMap) {
        this.schoolInfoMap = schoolInfoMap;
    }
    public Properties getFamilyProperties() {
        return familyProperties;
    }
    public void setFamilyProperties(Properties familyProperties) {
        this.familyProperties = familyProperties;
    }
    public float getAvg() {
        return avg;
    }
    public void setAvg(float avg) {
        this.avg = avg;
    }
    @Override
    public String toString() {
        return "SetClass [avg=" + avg + ", age=" + age + ", nameString=" + nameString + ",
        scoreClass=" + scoreClass + ", scorelist=" + scorelist + ", books=" + Arrays.toString
```

```
            (books) + ", gender=" + gender + ", schoolInfoMap="+ schoolInfoMap + ", family
            Properties=" + familyProperties + "]";
        }
    }
```

以上代码 SetClass 类有 avg、age 等普通基本数据类型的成员，还有成绩类 ScoreClass 对象、数组、集合等成员，并实现对应的 setter 方法。

(2) 创建 Spring 配置文件 beans.xml，在该文件中实现设置值方式的注入。

beans.xml

```
<?xml version="1.0" encoding="UTF-8"?>
<beans xmlns="http://www.springframework.org/schema/beans"
    xmlns:xsi="http://www.w3.org/2001/XMLSchema-instance"
    xmlns:p="http://www.springframework.org/schema/p"
    xsi:schemaLocation="http://www.springframework.org/schema/beans
    https://www.springframework.org/schema/beans/spring-beans.xsd">
    <bean id="score" class="demo4_3_1.ScoreClass">
        <property name="ch" value="89" />
        <property name="en" value="98" />
        <property name="math" value="100" />
    </bean>
    <bean id="set" class="demo4_3_1.SetClass" p:nameString="王新" p:age="18">
        <property name="scoreClass"   ref="score"/>
        <property name="avg" value="89.5" />
        <property name="books">
            <array>
              <value>"史记"</value>
              <value>"三国演义"</value>
              <value>"西游记"</value>
            </array>
        </property>
        <property name="familyProperties">
            <props>
                <prop key="brother">liming</prop>
                <prop key="sister">lucy</prop>
            </props>
        </property>
        <property name="schoolInfoMap">
            <map>
                <entry key="name" value="高新一中"/>
                <entry key="adress" value="电子二路"/>
```

```
            </map>
        </property>
        <property name="gender">
            <set>
                <value>男</value>
                <value>女</value>
            </set>
        </property>
        <property name="scorelist">
            <list>
                <value>98</value>
                <value>100</value>
                <value>101</value>
            </list>
        </property>
    </bean>
</beans>
```

以上代码中，使用<property>标签完成对 ScoreClass 普通基本数据类型的成员的依赖注入，使用<property>配合<array>、<props>、<map>、<set>、<list>标签对 SetClass 对象、数组和集合等类型成员的依赖注入，使用 p 前缀指定 nameString 属性的值。

(3) 创建测试类文件 Test.java，在该类中通过容器获取对应类的实例。

Test.java

```
package demo4_3_1;
import org.springframework.context.ApplicationContext;
import org.springframework.context.support.ClassPathXmlApplicationContext;
public class Test {
    public static void main(String arg[]) {
        ApplicationContext context = new ClassPathXmlApplicationContext("beans.xml");
        SetClass setClass=context.getBean("set", SetClass.class);
        System.out.println(setClass.toString());
    }
}
```

(4) 运行 Test.java 的 main 方法，从运行结果可以看出已经成功完成了对 SetClass 类的依赖注入，运行结果如下：

```
SetClass [avg=89.5, age=18, nameString=王新, scoreClass=demo4_2_3.ScoreClass@3059cbc,
    scorelist=[98, 100, 101], books=["史记", "三国演义", "西游记"], gender=[男, 女],
    schoolInfoMap={name=高新一中, adress=电子二路}, familyProperties={brother=liming,
    sister=lucy}]
```

多学一招：以上提供的 XML 配置片段中，xmlns:p 是一个命名空间的声明，它允许使用 p 前缀指定属性(property)的值。

4.3.2 Bean 的注解装配方式

基于 XML 的装配比较灵活，虽然文件的结构和语法相对直观，但是当配置变得复杂时，XML 文件可能会变得难以阅读和维护。Spring 提供基于 Annotation(注解)技术的装配方式。使用注解装配需要在 Spring 配置文件中使用 context 命名空间，并使用 annotation-config 开启注解处理器，具体方法如下：

```
<?xml version="1.0" encoding="UTF-8"?>
<beans xmlns="http://www.springframework.org/schema/beans"
    xmlns:xsi="http://www.w3.org/2001/XMLSchema-instance"
    xmlns:context="http://www.springframework.org/schema/context"
    xsi:schemaLocation="http://www.springframework.org/schema/beans
        https://www.springframework.org/schema/beans/spring-beans.xsd
        http://www.springframework.org/schema/context
        https://www.springframework.org/schema/context/spring-context.xsd">
        <context:annotation-config/>
</beans>
```

Spring 常见注解如下：

(1) @Value：注入外部属性值，例如使用@Value 给 catalog 参数注入外部变量 catalog.name 的值。具体代码如下：

```
public class Recommender {
    private final String catalog;
    public Recommender(@Value("${catalog.name}") String catalog) {
        this.catalog = catalog;
    }
}
```

(2) @Resource：对 Bean 的属性、setter 方法及构造方法进行标注，按照名称装配。如果没有指定 name 属性，则默认按照字段名称查找。如果注解写在 setter 方法上，则按照方法的参数名进行装配。例如，将名称为 myMovieFinder 的 Bean 注入到 movieFinder 参数，代码如下：

```
public class MovieLister {
    private MovieFinder movieFinder;
    @Resource(name="myMovieFinder")
    public void setMovieFinder(MovieFinder movieFinder) {
        this.movieFinder = movieFinder;
    }
}
```

(3) @Autowired：对 Bean 的属性、setter 方法及构造方法进行标注，按照类

型自动装配。默认情况下要求依赖对象必须存在，如果允许为 null，则可以设置 required 属性为 false，如@Autowired(required=false)。例如，将 CustomerDao 类型的 Bean 自动注入到 customerDao 参数，代码如下：

```
public class Recommender {
    private final CustomerDao customerDao;
    @Autowired
    public Recommender(CustomerDao customerDao) {
        this.customerDao = customerDao;
    }
}
```

(4) @Qualifier：与@Autowired 注解配合使用，会将@Autowired 默认的按类型装配修改为按 Bean 的实例名称装配，Bean 的实例名称将由注解@Qualifier 的参数指定。例如，将名称为 main 的 Bean 注入到 movieCatalog 参数，代码如下：

```
public class Movie {
    @Autowired
    @Qualifier("main")
    private Catalog movieCatalog;
}
```

(5) @Component：用于描述 Spring 中的 Bean，被@Component 注解的类会被 Spring 容器自动扫描并进行实例化，这样可以通过简单的注解来声明 Bean，可见@Component 相对前面几个注解简化了配置过程。例如，在类的定义上面添加@Component，对应 Bean 的 name 为 myclass，代码如下：

```
@Component("myclass")
public class Myclass {
    …
}
```

最后在 Spring 配置文件中按照以下方式开启扫描即可：

```
<context:component-scan base-package="org.example"/>
```

【案例 4-7】 类 SetClass 中有 avg、age 等普通基本数据类型的成员，还有成绩类对象、数组、集合等成员，利用 Spring 完成基于注解的依赖注入，具体步骤如下：

(1) 参照 4.1.2 节创建 Maven 项目，在 java 目录下创建包 demo4_3_2。

(2) 在 Resources 目录下创建属性文件 mydata.properties。

mydata.properties

```
test.mydata.ch=98
test.mydata.en=98
test.mydata.math=98
test.mydata.scorelist=108 120 98
test.mydata.books=englishbook \u4E09\u56FD\u6F14\u4E49    chinesebook
test.mydata.books=\u7537    \u5973
```

```
test.mydata.schoolInfoMap={adress:'\u5317\u5927\u8857', num:5}
test.mydata.familyProperties={sister:'\u4E3D\u534E', brother:'lili'}
```

(3) 创建 Bean 的类文件 ScoreClass.java 和 SetClass.Java，使用@Value 进行依赖注入，对 SetClass 的成员 scoreClass1 使用@Autowired 或者@Resource 进行依赖注入。

ScoreClass.java

```
package demo4_3_2;
import org.springframework.beans.factory.annotation.Value;
import org.springframework.context.annotation.Scope;
import org.springframework.stereotype.Component;
public class ScoreClass {
    @Value("99")
    int en;
    @Value("${test.mydata.ch}")
    int ch;
    @Value("${test.mydata.math}")
    int math;
}
```

SetClass.Java

```
package demo4_3_2;
import java.util.Arrays;
import java.util.List;
import java.util.Map;
import java.util.Properties;
import java.util.Set;
import org.springframework.beans.factory.annotation.Autowired;
import org.springframework.beans.factory.annotation.Qualifier;
import org.springframework.beans.factory.annotation.Value;
import org.springframework.stereotype.Component;
public class SetClass {
//注意 jdk11 以上已经不支持@resource 注解
@Autowired
@Qualifier("scoreClass1")
//以上两行注解也可以用  @Resource(name="score1")代替
ScoreClass scoreClass1;
@Value("78.5")
float avg;
@Value("98")
int age;
@Value("晓东")
```

```
    String nameString;
    @Value ("${test.mydata.scorelist}")
    List<Integer> scorelist;
    @Value("${test.mydata.books}")
    String[] books;
    @Value("${test.mydata.gender}")
    Set gender;
    @Value("#{${test.mydata.schoolInfoMap}}")
    Map<String, Object> schoolInfoMap;
    @Value("#{${test.mydata.familyProperties}}")
    Properties familyProperties;
    @Override
        public String toString() {
            return "SetClass [scoreClass1=" + scoreClass1 + ", avg=" + avg + ", age=" + age + ",
            nameString=" + nameString+ ", scorelist=" + scorelist + ", books=" + Arrays.toString
            (books) + ", gender=" + gender + ", schoolInfoMap="+ schoolInfoMap + ", family
            Properties=" + familyProperties + "]";
        }
    }
```

(4) 修改配置文件 beans.xml，在该配置文件中读取属性文件，增加了注解驱动，并配置了 3 个 Bean。

beans.xml

```
<?xml version="1.0" encoding="UTF-8"?>
<beans xmlns="http://www.springframework.org/schema/beans"
        xmlns:xsi="http://www.w3.org/2001/XMLSchema-instance"
        xmlns:context="http://www.springframework.org/schema/context"
        xsi:schemaLocation="
        http://www.springframework.org/schema/beans
        https://www.springframework.org/schema/beans/spring-beans.xsd
        http://www.springframework.org/schema/context
        https://www.springframework.org/schema/context/spring-context.xsd">
<!-- 读取 prop 文件，忽略解析不到的属性-->
    <context:property-placeholder location="classpath:mydata.properties" ignore-unresolvable
= "true"/>
<!—添加注解驱动-->
<context:annotation-config />
    <bean id="scoreClass1" class="demo4_3_2.ScoreClass"></bean>
    <bean id="scoreClass2" class="demo4_3_2.ScoreClass"></bean>
    <bean id="set" class="demo4_3_2.SetClass"></bean>
</beans>
```

(5) 创建测试类文件 Test.java，在该文件中获取 SetClass 类的实例。

Test.java

```
package demo4_3_1;
import org.springframework.context.ApplicationContext;
import org.springframework.context.support.ClassPathXmlApplicationContext;
public class Test {
    public static void main(String arg[]) {
        ApplicationContext context= new ClassPathXmlApplicationContext("beans.xml");
        SetClass set1=context.getBean("set", SetClass.class);
        System.out.println(set1.toString());
    }
}
```

(6) 运行 Test.java 的 main 方法，从运行结果可以看出使用注解方式成功完成对 SetClass 的依赖注入，运行结果如下：

```
SetClass [scoreClass1=demo4_2_4.ScoreClass@1f760b47, avg=78.5, age=98, nameString=晓东,
    scorelist=[10812098], books=[男  女], gender=[${test.mydata.gender}], schoolInfoMap=
    {adress=北大街, num=5}, familyProperties={brother=lili, sister=丽华}]
```

多学一招：注解方式需要显式地在 XML 文件中配置 Bean，因此可以通过扫描类路径而不需要使用 XML 来执行 Bean 注册，详见第 7 步以后操作。

(7) 添加@Component 注解，改为扫描方式，完成对 ScoreClass.java、SetClass.Java 和 applicationContext.xml 的修改。

ScoreClass.java 修改

```
…
import org.springframework.stereotype.Component;
//前面代码不变，添加组件注解，其中 scoreClass1 为 Bean 的 id
@Component("scoreClass1")
public class ScoreClass {
//后面代码不变
…
```

SetClass.Java 修改

```
…
import org.springframework.stereotype.Component;
//前面代码不变，添加组件注解，其中 set 为 Bean 的 id
@Component("set")
public class SetClass {
//后面代码不变
…
```

applicationContext.xml 修改

```
…
<context:annotation-config />
```

```
<!-- 前面代码不变，开启扫描 -->
<context:component-scan base-package="demo4_3_2"></context:component-scan>
</beans>
```

(8) 运行 Test.java 的 main 方法，发现扫描方式与普通注解方式结果相同。

多学一招：表达式 ${…}主要用于加载外部属性文件中的值；#{…}用于执行 SpEL 表达式，作用是通过 Spring 把值注入给某个属性；#{…}和 ${…}可以混合使用，但必须 #{}在外面而${}在里面。如@Value("#{'${jdbc.s_name}'}")。

为了更好地完成和维护项目，关于 Spring 依赖注入推荐以下方式：使用 XML 注入管理 Bean，使用注解完成属性注入；尽量不用扫描方式，因为扫描方式要增加类上的注解。

4.3.3 Bean 的自动装配方式

对于使用 XML 配置和定义 Bean，可以通过 autowire 属性来实现自动装配，不用写<constructor-arg>和<property>元素。Bean 的 autowire 属性取值如下：

(1) no：默认不自动装配，Bean 之间的依赖需要借助 ref 元素完成。

(2) byName：由属性名称自动装配，Spring 容器在配置文件中查找与之相同的 Bean 名称进行匹配。注意类中的 setter、getter 方法名称都要与属性名称对应。例如属性名为 master，则 setter 方法名应该为 setMaster。

(3) byType：由属性类型自动装配，Spring 根据类型匹配对应 Bean，如果有多个满足该条件的 Bean，则抛出异常。

(4) constructor：类似于 byType，但该类型适用于构造函数的参数装配。

【案例 4-8】本案例中，使用 Spring 配置 3 个 Bean，分别使用 byName、byType 和 constructor 进行自动装配，具体实现步骤如下：

(1) 参照 4.1.2 节完成步骤 1～8 后，在 java 目录下创建包 demo4_3_3。

(2) 在该包下创建 4 个类文件，分别是 ScoreClass.Java、AutoWireSetClass1.java、AutoWireSetClass2.java、AutoWireConstructorClass.java，最后 3 个类都依赖于第 1 个类即 ScoreClass 类。

ScoreClass.Java

```
package demo4_3_3;
public class ScoreClass {
    public int getEn() {
        return en;
    }
    public void setEn(int en) {
        this.en = en;
    }
    public int getCh() {
        return ch;
```

```
        }
        public void setCh(int ch) {
            this.ch = ch;
        }
        public int getMath() {
            return math;
        }
        public void setMath(int math) {
            this.math = math;
        }
        int en;
        int ch;
        int math;
}
```

AutoWireSetClass1.java

```
package demo4_3_3;
public class AutoWireSetClass1 {
    //要和 bean 名称一致
    ScoreClass scoreClass;
    public ScoreClass getScoreClass() {
        return scoreClass;
    }
    public void setScoreClass(ScoreClass scoreClass1) {
        this.scoreClass = scoreClass1;
    }
    @Override
    public String toString() {
        return "AutoWireClass1 [scoreClass=" + scoreClass + "]";
    }
}
```

AutoWireSetClass2.java

```
package demo4_3_3;
public class AutoWireSetClass2 {
ScoreClass scoreClass2;
    public ScoreClass getScoreClass() {
        return scoreClass2;
    }
    public void setScoreClass(ScoreClass scoreClass) {
        this.scoreClass2 = scoreClass;
```

```
        }
        @Override
        public String toString() {
            return "AutoWireClass2 [scoreClass=" + scoreClass2 + "]";
        }
    }
```

AutoWireConstructorClass.java

```
package demo4_3_3;
public class AutoWireConstructorClass {
    ScoreClass scoreClass;
    public AutoWireConstructorClass(ScoreClass scoreClass) {
        super();
        this.scoreClass = scoreClass;
    }
    @Override
      public String toString() {
        return "AutoWireClass [scoreClass=" + scoreClass + "]";
      }
}
```

(3) 创建 Spring 配置文件 beans.xml，对 3 个 bean 分别使用 byName、byType 和 constructor 进行自动装配。

beans.xml

```
<?xml version="1.0" encoding="UTF-8"?>
<beans xmlns="http://www.springframework.org/schema/beans"
xmlns:xsi="http://www.w3.org/2001/XMLSchema-instance"
xmlns:p="http://www.springframework.org/schema/p"
xsi:schemaLocation="http://www.springframework.org/schema/beans
        https://www.springframework.org/schema/beans/spring-beans.xsd">
    <bean id="scoreClass" class="demo4_3_3.ScoreClass">
        <property name="ch" value="89" />
        <property name="en" value="98" />
        <property name="math" value="100" />
    </bean>
    <bean id="autoset1" class="demo4_3_3.AutoWireSetClass1"
autowire="byName">
    </bean>
    <bean id="autoset2" class="demo4_3_3.AutoWireSetClass2"
autowire="byType">
```

```
    </bean>
    <bean id="autoconstructor" class="demo4_3_3.AutoWireConstructorClass"
autowire="constructor">
    </bean>
</beans>
```

(4) 创建测试类文件 Test.java，该文件中使用 Spring 获取对应类的实例。

Test.java

```
package demo4_3_3;
import org.springframework.context.ApplicationContext;
import org.springframework.context.support.ClassPathXmlApplicationContext;
public class Test {
public static void main(String arg[]) {
        ApplicationContext context = new ClassPathXmlApplicationContext("beans.xml");
        AutoWireSetClass1 autoset1=context.getBean("autoset1", AutoWireSetClass1.class);
        System.out.println(autoset1.toString());
        AutoWireSetClass2 autoset2=context.getBean("autoset2", AutoWireSetClass2.class);
        System.out.println(autoset2.toString());
AutoWireConstructorClass autoWireCon=context.getBean("autoconstructor",
AutoWireConstructorClass.class);
        System.out.println(autoWireCon.toString());
    }
}
```

(5) 运行 Test.java 的 main 方法，结果如下：

```
AutoWireClass1 [scoreClass=demo4_3_3.ScoreClass@73700b80]
AutoWireClass2 [scoreClass=demo4_3_3.ScoreClass@73700b80]
AutoWireClass [scoreClass=demo4_3_3.ScoreClass@73700b80]
```

由以上可知，3 个 Bean 自动装配完成对 ScoreClass 类的依赖注入。

4.4　Spring AOP

4.4.1　Spring AOP 简介

通常情况下，将跨越应用程序多个模块或组件，与应用程序的核心业务逻辑不直接相关，但却是实现完整功能所必需的，如日志、安全、事务管理等通用任务，称之为横切关注点。在传统的面向对象编程(Object-Oriented Programming，OOP)中，横切关注点通常会被混杂到业务逻辑中，导致代码混乱而难以维护。

面向切面编程(Aspect-Oriented Programming，AOP)通过引入“切面”(Aspects)

的概念，将横切关注点从业务逻辑中分离出来，允许开发者在不修改业务代码的情况下，通过切面对通用任务模块化和重用处理，使得这些横切关注点可以被统一地管理和应用于对象的不同部分，从而提高代码的模块化和可维护性。Spring 提供对 AOP 的支持，Spring AOP 是 Spring 框架的一个重要组成部分。

AOP 相关术语如下：

(1) Aspect(切面)：表示跨越多个对象和类的横切功能，例如日志、安全检查、事务管理等。应用程序可以拥有任意数量的切面。

(2) Join point(连接点)：表示在程序执行过程中的某个特定时刻，如某个异常处理后、某个方法执行前或执行后。

(3) Pointcut(切入点)：表示被通知所拦截的方法或者类。

(4) Advice(通知/增强处理)：表示切面在特定的连接点上应该执行的操作，即在切入点实际被调用的代码。Spring AOP 通知类型如表 4-2 所示。

表 4-2 通知类型

通　知	描　　述
前置通知	在一个方法执行之前执行，常用来进行准备工作
后置通知	在一个方法执行之后执行，无论正常返回或异常返回都要运行该通知，常用来处理方法返回值后的逻辑
返回后通知	在一个方法执行之后且成功完成时进行一些操作，如验证返回值
抛出异常后通知	在一个方法执行之后且抛出异常时执行的操作
环绕通知	最强大的通知类型，可以在方法执行之前和之后做任何事情，甚至可以决定是否执行该方法

(5) Target object(目标对象)：被一个或者多个切面所通知的对象，Spring AOP 中，这个对象总是一个被代理的对象。

(6) Weaving(织入)：表示 Spring AOP 在运行时将切面应用于对象的过程，即把横切功能代码插入到被代理对象上，从而生成新的对象的过程。

4.4.2 使用 AspectJ 的 Spring AOP 实现

AspectJ 是一个基于 Java 面向切面编程(AOP)的框架，最新的 Spring 框架推荐使用 AspectJ 开发 AOP。AspectJ 提供一系列注解来简化 AOP 的定义，包括：

- @Pointcut：定义一个切入点表达式。
- @Before：在目标方法执行之前执行的通知。
- @AfterReturning：在目标方法正常返回后执行的通知。
- @AfterThrowing：在目标方法抛出异常后执行的通知。
- @After：在目标方法执行之后执行的通知，无论是否抛出异常。
- @Around：围绕目标方法执行的通知。

通过使用 AspectJ 的注解，可以将 AOP 的定义与业务逻辑分离，使代码更加

清晰和模块化。下面将通过一个中介租房业务逻辑案例，演示如何使用 AspectJ 实现 Spring AOP。

【案例 4-9】 本案例使用注解 Aspect 方式定义切面和通知，完成其他通用任务的模块化和重用处理，实现步骤如下：

(1) 参照 4.1.2 节创建 Maven 项目，在 java 目录下创建包 demo4_3。

(2) 在该包下创建接口类文件 RentHouse.java。

RentHouse.java

```
package demo4_3;
import java.util.Map;
public interface RentHouse {
    public abstract Map<String, String> rent();
}
```

(3) 在 pom.xml 中增加 spring-aspects 依赖。

pom.xml 修改部分

```
<dependency>
        <groupId>org.springframework</groupId>
        <artifactId>spring-aspects</artifactId>
        <version>5.3.24</version>
    </dependency>
```

(4) 创建被代理类文件 Host.java，在该类中实现被代理的方法 rent。

Host.java

```
package demo4_3;
import java.util.HashMap;
import java.util.Map;
import org.springframework.beans.factory.annotation.Value;
public class Host implements RentHouse{
    @Value("房东老王")
    String hostNameString ;
    @Value("西安市碑林区兴庆路 xxxx 号")
    String houseAdressString;
    @Override
    public Map<String, String> rent() {
        Map hostMap=new HashMap<String, String>();
        hostMap.put("hostname", hostNameString);
        hostMap.put("housename", houseAdressString);
        System.out.println("房东完成房屋出租：房东="+hostMap.get("hostname")+"  出租=
                    "+hostMap.get("housename"));
        //异常测试
        //int i=10/0;
```

```
        return hostMap;
    }
}
```

(5) 创建切面类文件 MyAspect.java。

在该类中，使用@Pointcut("execution(* demo4_3.*.*(..))")中定义了一个空方法 myPointCut 来命名切入点，其中第 1 个*代表任意返回值，第 2 个*代表任何类，第 3 个*代表任何方法；使用@Before、@After、@Around 和@AfterReturning 分别定义了前置通知、后置通知、环绕通知和返回通知；使用@AfterThrowing 注解的 throwing 属性接收 eThrowable 类型的异常。

MyAspect.java

```
package demo4_3;
import org.springframework.beans.factory.annotation.Value;
import org.aspectj.lang.JoinPoint;
import org.aspectj.lang.ProceedingJoinPoint;
import org.aspectj.lang.annotation.After;
import org.aspectj.lang.annotation.AfterReturning;
import org.aspectj.lang.annotation.AfterThrowing;
import org.aspectj.lang.annotation.Around;
import org.aspectj.lang.annotation.Aspect;
import org.aspectj.lang.annotation.Before;
import org.aspectj.lang.annotation.Pointcut;
@Aspect
public class MyAspect {
    @Value("代理商甲")
    String proxyNameString;
    @Value("客户王先生")
    String guestNamString;
    //定义一个空方法来命名切入点
    @Pointcut("execution(* demo4_3.*.*(..))")
    public void myPointCut() {
    }
    //被代理的方法执行前
    @Before("myPointCut()")
    public void seeHouse(JoinPoint joinpoint) {
        System.out.println(proxyNameString + "带客户" + guestNamString + "看房");
    }
    //被代理的方法执行后
    @After("myPointCut()")
    public void getFare(JoinPoint joinpoint) {
```

```
        System.out.println(proxyNameString + "收客户" + guestNamString + "房租");
    }
    /* 环绕通知
     * ProceedingJoinPoint 是 JoinPoint 子接口，表示可以执行目标方法
     * 1.必须是 Object 类型的返回值
     * 2.必须接收一个参数，类型为 ProceedingJoinPoint
     * 3.必须 throws Throwable
     */
    @Around("myPointCut()")
    public void communicate(ProceedingJoinPoint joinpoint) {
        System.out.println(joinpoint.getSignature().getName() + "执行前，沟通");
        try {
        //执行目标方法
            joinpoint.proceed();
        }
        catch (Throwable e) {
            e.printStackTrace();
        }
        System.out.println(joinpoint.getSignature().getName() + "执行后，沟通");
    }
    //执行完返回后
    @AfterReturning(value = "myPointCut()", returning = "returnObject")
    public void haveresult(JoinPoint joinPoint, Object returnObject) {
        System.out.println(joinPoint.getSignature().getName() + "执行完返回后得到返回值" +
                        returnObject.toString());
    }
    //出错后
    @AfterThrowing(value = "myPointCut()", throwing = "eThrowable")
    public void haveError(JoinPoint joinPoint, Throwable eThrowable) {
        System.out.println(joinPoint.getSignature().getName() + "执行完返回后发生错误" +
                        eThrowable.getMessage());
    }
}
```

(6) 修改配置文件 applicationContext.xml，增加 aop 命名空间和配置，添加目标类、切面类配置以及 AspectJ 注解驱动。

applicationContext.xml

```
<?xml version="1.0" encoding="UTF-8"?>
<beans xmlns="http://www.springframework.org/schema/beans"
    xmlns:xsi="http://www.w3.org/2001/XMLSchema-instance"
```

```
        xmlns:context="http://www.springframework.org/schema/context"
        xmlns:aop="http://www.springframework.org/schema/aop"
        xsi:schemaLocation="http://www.springframework.org/schema/beans
        https://www.springframework.org/schema/beans/spring-beans.xsd
        http://www.springframework.org/schema/context
        https://www.springframework.org/schema/context/spring-context.xsd
         http://www.springframework.org/schema/aop
        https://www.springframework.org/schema/aop/spring-aop.xsd">
<context:annotation-config/>
    <!-- 目标类 -->
    <bean id="host" class="demo4_3.Host"></bean>
    <!-- 切面类 -->
    <bean id="myaspect" class="demo4_3.MyAspect"></bean>
    <!--   声明基于注解的 aspectj   -->
    <aop:aspectj-autoproxy/>
</beans>
```

(7) 创建测试类文件 Test.java，实现获取代理类实例。

Test.java

```
package demo4_3;
import org.springframework.context.ApplicationContext;
import org.springframework.context.support.ClassPathXmlApplicationContext;
public class Test {
public static void main(String[] a) {
        ApplicationContext context=new ClassPathXmlApplicationContext("beans.xml");
        RentHouse   rentHouse=context.getBean("proxy", RentHouse.class);
        rentHouse.rent();
    }
}
```

(8) 运行 Test.java 的 main 方法，最终结果如图 4-14 所示。

```
Run:  Test
"C:\Program Files\Java\jdk1.8.0_211\bin\java.exe" ...
rent执行前，沟通
代理商甲带客户客户王先生看房
房东完成房屋出租：房东=房东老王    出租=西安市碑林区兴庆路xxxx号
rent执行完返回后得到返回值{hostname=房东老王, housename=西安市碑林区兴庆路xxxx号}
代理商甲收客户客户王先生房租
rent执行后，沟通
```

图 4-14 运行结果

由以上结果可以看出，在被代理类的 rent 方法执行前后，切面类中的方法已经成功执行；放开 Host.java 中“int i=10/0;”语句的注释，重新运行测试类的 main

方法，结果如图 4-15 所示，发现异常通知已经执行。

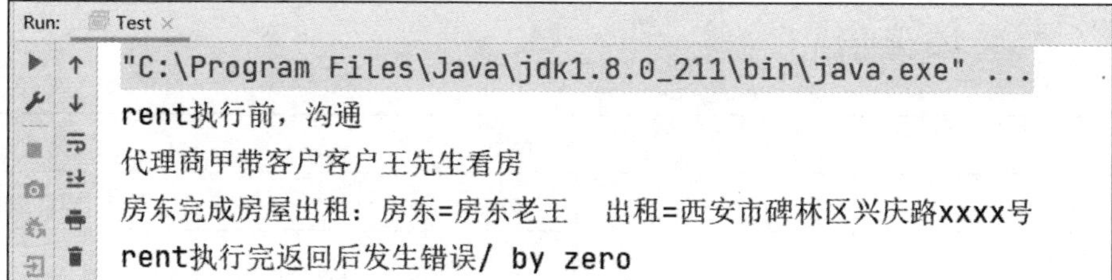

图 4-15　异常通知结果

由以上可以看出，使用注解 Aspect 方式没有修改原有的 Host.java 中租房业务逻辑代码，通过注解定义切面和通知，完成其他通用任务的模块化和重用处理，最终完成了 Spring AOP 的开发。

强化练习

习题答案

1. 在 Spring 框架中，(　　)不是 Bean 的作用域。
A. singleton　　B. prototype　　C. request　　D. application
2. 在 Spring 框架中，(　　)类负责管理 Bean 的生命周期。
A. BeanFactory　　B. ApplicationContext
C. BeanWrapper　　D. BeanBuilder
3. 在 Spring 框架中，(　　)注解不能用于字段注入。
A. @Service　　B. @Autowired　　C. @Qualifier　　D. @Resource
4. 在 Spring 框架中，(　　)不是 AOP 的核心概念。
A. 切点　　B. 通知　　C. 代理　　D. 数据源
5. 在 Spring 框架中，(　　)注解用于声明一个切点。
A. @Pointcut　　B. @Before　　C. @After　　D. @Around
6. 在 Spring 框架中，(　　)不是常用的 AOP 通知类型。
A. Before　　B. After　　C. Around　　D. Transaction
7. 在 Spring 框架中，(　　)不是常用的依赖注入注解。
A. @Autowired　　B. @Resource　　C. @Inject　　D. @Qualifier

进一步学习建议

学习 Spring 基础之后，可以进一步学习以下内容：
(1) 学习 Spring JDBC 以及事务管理。
(2) 学习 MyBatis 与 Spring 集成进行数据库操作。
(3) 理解 Spring MVC 的工作流程，包括请求映射、视图解析和模型管理。
(4) 学习使用 Spring Test 框架进行单元测试和集成测试。

考核评价

<table>
<tr><td colspan="5">考核评价表</td></tr>
<tr><td colspan="2">姓名</td><td></td><td>班　　级</td><td></td></tr>
<tr><td colspan="2">学号</td><td></td><td>考评时间</td><td></td></tr>
<tr><td colspan="2">评价主题及总分</td><td colspan="2">评价内容及分数</td><td>评分</td></tr>
<tr><td rowspan="3">1</td><td rowspan="3">知识考核
(30)</td><td colspan="2">阐述 Spring 框架的主要特点(10 分)</td><td></td></tr>
<tr><td colspan="2">举例说明作用域 singleton 和 prototype 的区别(10 分)</td><td></td></tr>
<tr><td colspan="2">阐述对 AOP 切面、连接点、切入点和通知的理解(10 分)</td><td></td></tr>
<tr><td rowspan="3">2</td><td rowspan="3">技能考核
(40)</td><td colspan="2">使用 XML 的构造器注入和设置值注入方式配置 Bean(10 分)</td><td></td></tr>
<tr><td colspan="2">掌握注解方式配置 Spring 框架的 Bean(20 分)</td><td></td></tr>
<tr><td colspan="2">熟练使用 Spring 框架实现 AOP 编程(10 分)</td><td></td></tr>
<tr><td rowspan="3">3</td><td rowspan="3">思政考核
(30)</td><td colspan="2">展示自主学习的知识、技能以及解决的问题(10 分)</td><td rowspan="3"></td></tr>
<tr><td colspan="2">讲解如何与他人共同工作，分享知识与经验(10 分)</td></tr>
<tr><td colspan="2">能总结、理解、编写可读、可维护且高效的代码(10 分)</td></tr>
<tr><td colspan="4">评语：</td><td>汇总：</td></tr>
</table>

第 5 章　Spring MVC 框架

学习目标

目标类型	目 标 描 述
知识目标	• 理解 Spring MVC 的工作流程 • 理解 Spring MVC 的核心组件及其作用 • 掌握 Spring MVC 接收请求数据的常见方式 • 掌握 Spring MVC 应用程序处理 JSON 格式的请求和响应的方法
技能目标	• 能够配置 Spring MVC 的核心组件，如控制器、视图解析器等 • 能够编写用于接收请求数据的控制器方法，实现绑定不同数据类型的形参，获取 HTTP 请求的查询数据或者表单数据 • 能够使用 Jackson 或 Gson 库实现 JSON 数据的转换 • 能够使用 Spring MVC 的 RESTful 支持编写代码
思政目标	• 树立正确的价值观，认识到技术进步对社会发展的积极作用，助力我国互联网产业发展 • 理解并实践开放共享的理念，建立社会主义核心价值观 • 能够将理论与实践相结合，解决实际编程问题，培养解决复杂问题的能力和创新精神

知识技能储备

5.1　Spring MVC 快速上手

5.1.1　Spring MVC 介绍

Spring MVC 即 Spring Web MVC，是基于 Servlet API 构建的 Web 框架，从一开始就被包含在 Spring Framework 中。Spring Web MVC 这个正式名称来源于其源模块的名称(spring-webmvc)。MVC 设计模式是一种架构模式，其核心思想是将数据

和显示逻辑分离，使得同一套数据模型可以配合不同的视图呈现，提高应用程序的灵活性和可维护性。它包括模型(Model)、视图(View)和控制器(Controller) 3 个主要部分。模型与数据库紧密相关，负责数据处理和业务逻辑；视图负责展示数据，常指用户看到的网页界面；而控制器则负责处理用户的输入并调用模型的逻辑来响应这些输入。

在 Web 应用开发中，当用户通过浏览器访问一个 URL 时，Web 服务器将其视为一个请求，控制器负责解析这个请求，执行必要的业务逻辑，并将最终结果呈现给用户，这个结果通常是一个视图。在这个过程中，用户输入的数据会被封装到一个实体类中，这个实体类就是模型的一部分。模型不仅负责存储数据，还负责数据的验证、检索和更新；控制器充当用户界面和数据模型层之间的桥梁，它处理用户的指令，决定如何响应用户的行为；视图则根据控制器提供的数据来生成用户看到的页面。

Spring MVC 的工作流程可以概括为以下几个步骤：

(1) 客户端发起请求：客户端(通常是浏览器)向 Spring MVC 应用发送一个 HTTP 请求。

(2) 前端控制器(DispatcherServlet)处理： DispatcherServlet 接收请求，并查询处理器映射器(HandlerMapping)来确定哪个控制器处理该请求。

(3) 处理器映射器(HandlerMapping)解析请求：HandlerMapping 负责将请求映射到具体的控制器处理方法，通常使用注解(如@RequestMapping)来配置映射。

(4) 控制器处理请求：Controller 接收请求并处理业务逻辑。

(5) ModelAndView 对象创建：业务处理完成后，Controller 会创建一个 ModelAndView 对象，其中 Model 用于存储请求的数据，View 用于存储返回的视图名称。

(6) 视图渲染：视图解析器(ViewResolver)根据 View 名称解析出真正的视图 View。

(7) 返回响应：渲染好的视图通过 DispatcherServlet 返回给客户端。

5.1.2 Spring MVC 入门指南

Spring MVC 入门

为了帮助读者快速掌握 Spring MVC 的使用，下面通过一个入门程序来演示编程过程，具体步骤如下：

(1) 参照 4.1.2 节创建一个名称为 FirstSpringMVC 的 Maven 项目，完成步骤(1)～(5)后增加以下项目依赖：

```
<dependency>
    <groupId>org.springframework</groupId>
    <artifactId>spring-context-support</artifactId>
    <version>5.3.24</version>
</dependency>
<dependency>
```

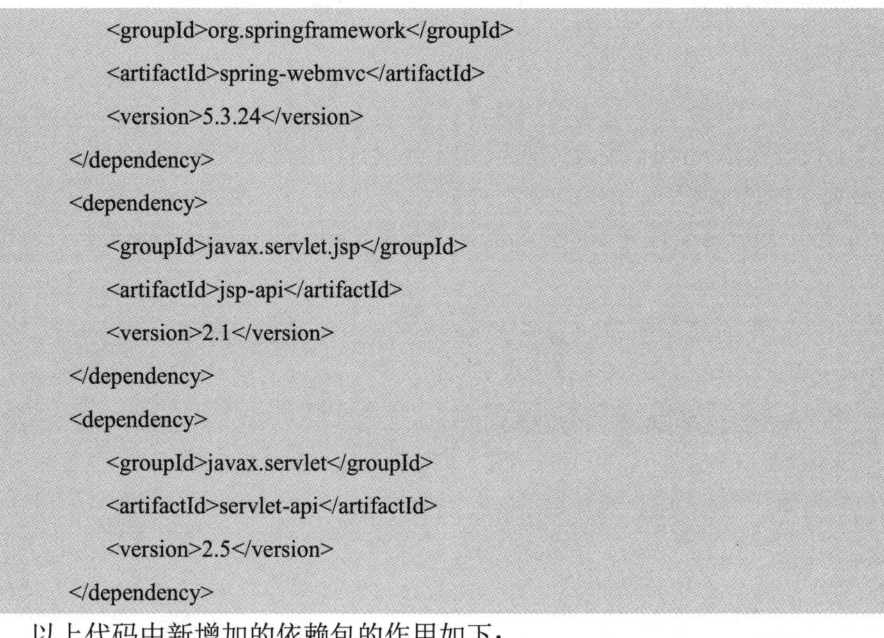

```
        <groupId>org.springframework</groupId>
        <artifactId>spring-webmvc</artifactId>
        <version>5.3.24</version>
    </dependency>
    <dependency>
        <groupId>javax.servlet.jsp</groupId>
        <artifactId>jsp-api</artifactId>
        <version>2.1</version>
    </dependency>
    <dependency>
        <groupId>javax.servlet</groupId>
        <artifactId>servlet-api</artifactId>
        <version>2.5</version>
    </dependency>
```

以上代码中新增加的依赖包的作用如下：

① spring-context-support：这是 Spring 框架的一个模块，提供了对 Spring 应用程序上下文的支持。

② spring-webmvc：这是 Spring MVC 框架的核心模块，提供了基于 Java 的 Web 应用程序开发所需的功能。

③ jsp-api：这是 Java Server Pages(JSP)规范的 API，提供了在 Java Web 应用程序中使用 JSP 页面所需的接口和类。

④ servlet-api：这是 Java Servlet API 的一部分，提供了在 Java Web 应用程序中处理 HTTP 请求和响应的接口和类。

(2) 在已创建项目的 java 文件夹中创建一个包 com.controller，在 com.controller 包下创建一个 Java 类 HelloController.java，该类继承控制器类。

HelloController.java

```
package com.controller;
import javax.servlet.http.HttpServletRequest;
import javax.servlet.http.HttpServletResponse;
import org.springframework.web.servlet.ModelAndView;
import org.springframework.web.servlet.mvc.Controller;
public class HelloController implements Controller {
@Override
    public ModelAndView handleRequest(HttpServletRequest arg0, HttpServletResponse arg1)
    throws Exception {
        ModelAndView mView=new ModelAndView();
        mView.addObject("msg", "hello springmvc");
        mView.setViewName("/WEB-INF/jsp/hello.jsp");
```

```
            return mView;
        }
    }
```

以上代码为 Controller 接口实现类，用于处理 HTTP 请求并返回一个包含“hello springmvc”消息的视图。

(3) 在 resources 文件夹下创建 Spring 的配置文件 applicationContext.xml，如图 5-1 所示，并输入代码。

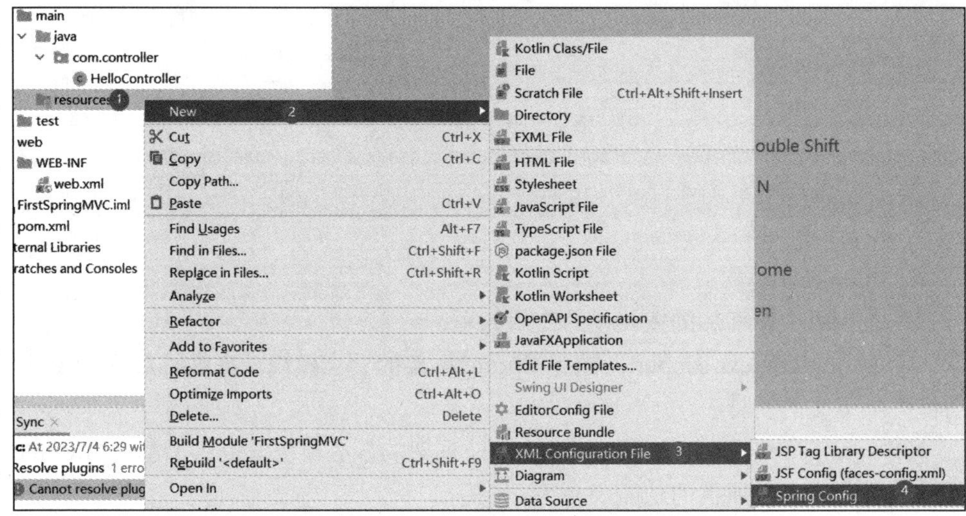

图 5-1　创建 Spring 配置文件

applicationContext.xml

```
<?xml version="1.0" encoding="UTF-8"?>
<beans xmlns="http://www.springframework.org/schema/beans"
xmlns:xsi="http://www.w3.org/2001/XMLSchema-instance"
xmlns:context="http://www.springframework.org/schema/context"
xsi:schemaLocation="
        http://www.springframework.org/schema/beans
        https://www.springframework.org/schema/beans/spring-beans.xsd
        http://www.springframework.org/schema/context
        https://www.springframework.org/schema/context/spring-context.xsd">
        <bean name="/HelloController" class="com.controller.HelloController"/>
</beans>
```

以上代码定义了一名称为“/HelloController”的 Bean，其类名称属性 class 对应为上一步创建的控制器类的完整路径。

(4) 添加 web.xml 并修改位置，web.xml 文件用于配置 Java 企业版(Java EE)应用程序的部署，其添加和修改位置方法如下：

① 单击 File→Project Structure 后添加 web.xml 文件，如图 5-2 所示。

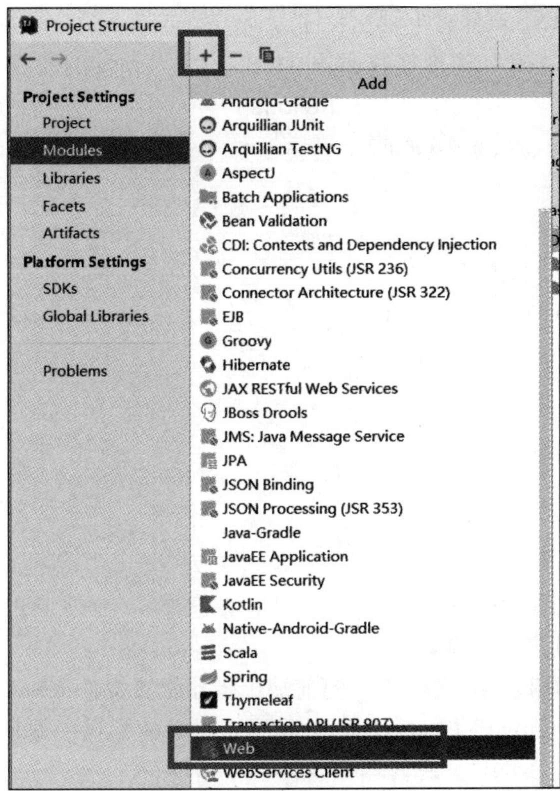

图 5-2　创建 web.xml

② 修改 web.xml 的位置为 src\main\web\WEB-INF\web.xml，如图 5-3 所示。

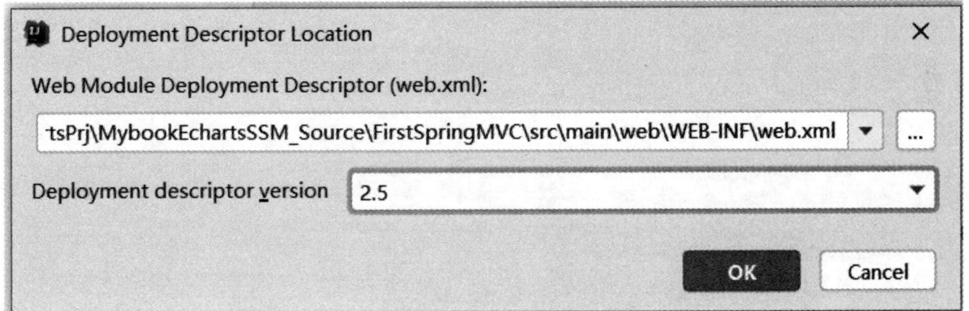

图 5-3　修改 web.xml 路径

③ 修改 web 目录路径为 src\main\web，如图 5-4 所示。

图 5-4　修改 web 目录路径

(5) 在 /WEB-INF 文件夹下创建名称为 jsp 的子文件夹，在这个子文件夹下创建一个视图文件 hello.jsp，实现接收服务端回传的名为 msg 的消息。

hello.jsp

```
<%@ page language="java" contentType="text/html; charset=UTF-8" pageEncoding="UTF-8" %>
<html>
<head>
    <meta charset="UTF-8">
    <title>my first springmvc</title>
</head>
<body>
<h1>${msg}</h1>
<p>你好</p>
</body>
</html>
```

以上代码使用 ${msg}作为占位符，用于动态插入一个变量的值，这个变量是从服务器端传递过来的数据。

(6) 修改在 /WEB-INF 文件夹下的前端控制器配置文件 web.xml，该文件定义了一个名为“springmvc”的 servlet，并指定了相关的配置参数和映射规则。

多学一招：WEB-INF 里的内容只能由服务器级别访问，客户端级别不能访问。

web.xml

```
<?xml version="1.0" encoding="UTF-8"?>
<web-app xmlns="http://java.sun.com/xml/ns/javaee"
        xmlns:xsi="http://www.w3.org/2001/XMLSchema-instance"
        xsi:schemaLocation="http://java.sun.com/xml/ns/javaee
            http://java.sun.com/xml/ns/javaee/web-app_2_5.xsd"
           version="2.5">
    <servlet>
<!-- 配置前端控制器 -->
        <servlet-name>springmvc</servlet-name>
        <servlet-class>org.springframework.web.servlet.DispatcherServlet</servlet-class>
<!--  初始化加载配置文件 -->
        <init-param>
            <param-name>contextConfigLocation</param-name>
            <param-value>classpath:applicationContext.xml</param-value>
        </init-param>
<!-- 容器启动时立即加载 servlet -->
        <load-on-startup>1</load-on-startup>
    </servlet>
<!--  拦截所有的 URL 交前端控制器处理 -->
```

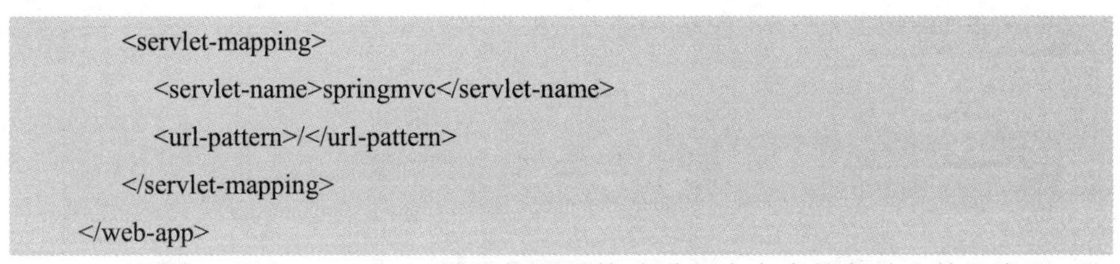

```
        <servlet-mapping>
            <servlet-name>springmvc</servlet-name>
            <url-pattern>/</url-pattern>
        </servlet-mapping>
    </web-app>
```

(7) 配置 Artifacts。Artifacts 通常指项目构建过程中产生的各种文件和资源，这些可以是编译后的类文件、配置文件、静态资源等。配置 Artifacts 如图 5-5 所示，选择 Artifacts 部署方式，将模块所有文件和资源直接复制到应用服务器的特定目录。

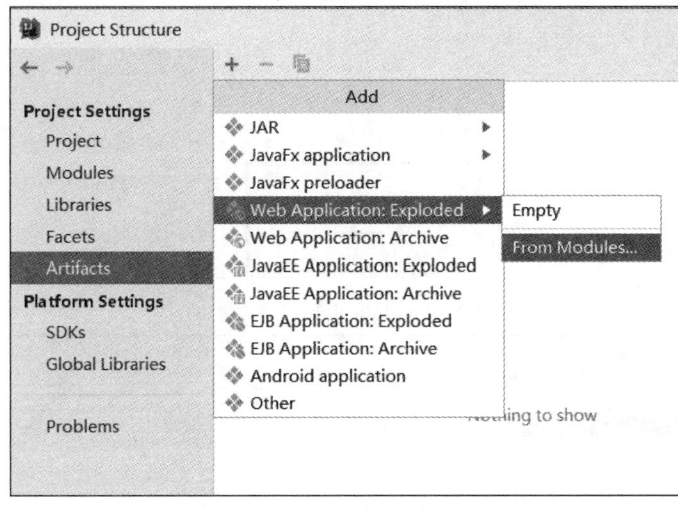

图 5-5　配置 Artifacts

(8) 配置 Tomcat。Tomcat 是一个用于构建和部署 Java Web 应用程序的开源软件。配置 Tomcat 时，首先按照如图 5-6 所示的方式添加配置，并按照如图 5-7 所示选择“Tomcat Server”的“Local”选项添加 Tomcat 服务器，选择本地安装的 Tomcat 目录；然后选择 Deployment 选项卡，按照如图 5-8 所示的方式选择 Artifact，向 Tomcat 配置 Artifacts；最后选择 Server 选项卡，按照如图 5-9 所示添加 URL 路径完成配置。

图 5-6　添加配置

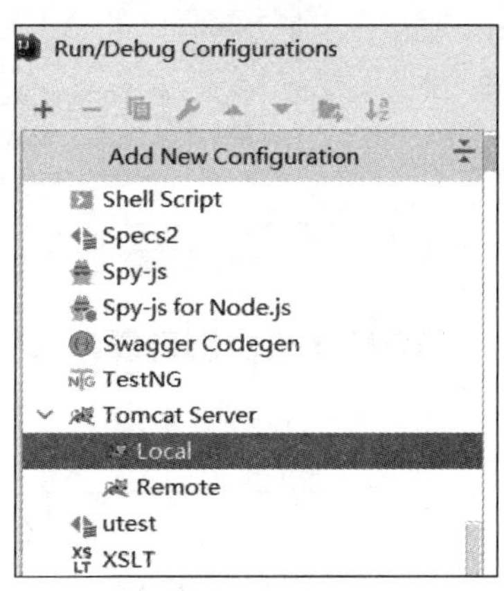

图 5-7　添加 Tomcat 服务器

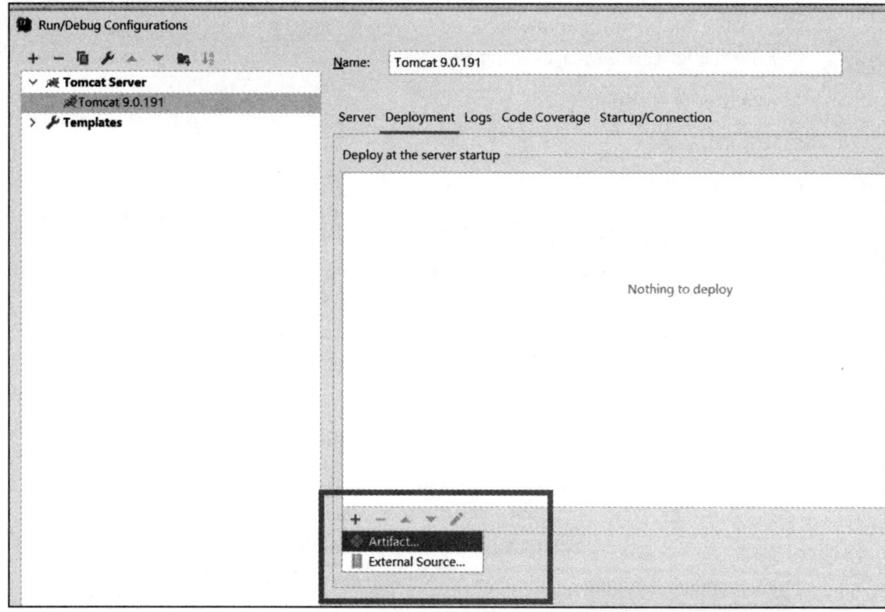

图 5-8　向 Tomcat 配置 Artifacts

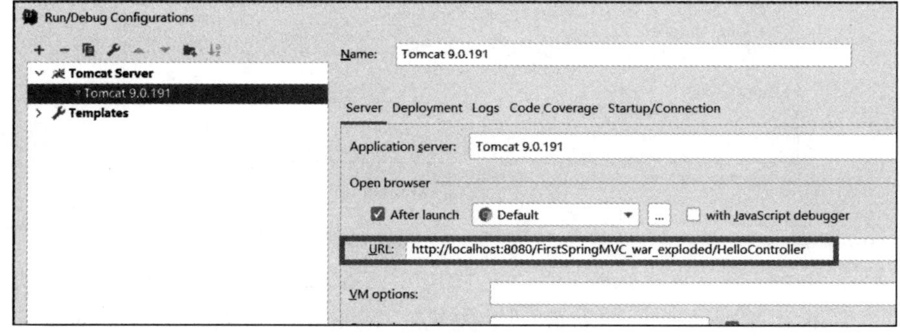

图 5-9　添加 URL 路径

(9) 单击如图 5-10 所示的运行按钮，运行 Tomcat 服务，浏览器自动打开，出现如图 5-11 所示的结果。

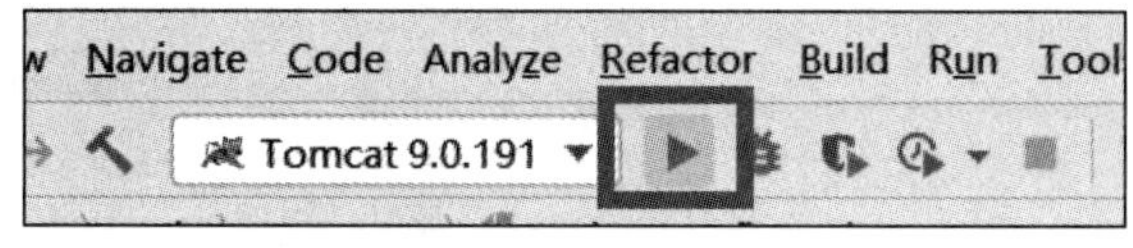

图 5-10　运行按钮

图 5-11　运行结果

由以上结果可以看出 Spring MVC 项目正常运行，并向客户端返回一个包含“hello springmvc”消息的视图。

5.2　Spring MVC 核心组件

5.2.1　DispatcherServlet

DispatcherServlet 也称为前端控制器，在 Spring MVC 框架中扮演着核心角色，它的主要职责是拦截前端请求，如同 Servlet 一样，DispatcherServlet 需要在 web.xml 中进行配置。

web.xml

```
<?xml version="1.0" encoding="UTF-8"?>
<web-app xmlns="http://java.sun.com/xml/ns/javaee"
        xmlns:xsi="http://www.w3.org/2001/XMLSchema-instance"
        xsi:schemaLocation="http://java.sun.com/xml/ns/javaee
            http://java.sun.com/xml/ns/javaee/web-app_2_5.xsd"
        version="2.5">
    <servlet>
        <servlet-name>dispatcherServlet</servlet-name>
        <servlet-class>org.springframework.web.servlet.DispatcherServlet</servlet-class>
        <init-param>
            <param-name>contextConfigLocation</param-name>
            <param-value>classpath:springMVC.xml</param-value>
        </init-param>
        <load-on-startup>1</load-on-startup>
    </servlet>
    <servlet-mapping>
        <servlet-name>dispatcherServlet</servlet-name>
        <url-pattern>/</url-pattern>
    </servlet-mapping>
</web-app>
```

以上代码中的<servlet>标签定义了一个名为 dispatcherServlet 的 Servlet，对应类名是 org.springframework.web.servlet.DispatcherServlet。具体配置解释如下：

(1) <init-param>标签：定义 DispatcherServlet 的初始化参数，其中参数 contextConfigLocation 参数显式指定 Spring MVC 的配置文件位于类路径下，且配置文件名为 springMVC.xml。若没有明确指定该路径和文件名，Spring MVC 初始化时会自动从 WEB-INF 目录下查找 Spring MVC 的配置文件，配置文件的默认命名规则为{servlet-name}-servlet.xml，如 springMVC-servlet.xml。

(2) <load-on-startup>1</load-on-startup>标签：指定 DispatcherServlet 应该在 Web 应用程序启动时被加载，这里的值 1 表示应该在 Servlet 容器启动时最早加载。load-on-startup 元素的取值必须是一个整数，当取值小于 0 或者没有指定时，表示

容器在该 Servlet 被首次请求时才会被加载，当取值大于 0 或等于 0 时，表示容器在启动时就加载并初始化该 Servlet，取值越小，加载优先级越高。

以上代码中的<servlet-mapping>标签定义 DispatcherServlet 的 URL 映射，这意味着 DispatcherServlet 将处理所有指向 / 的请求。这里的<url-pattern>/</url-pattern>表示 DispatcherServlet 将拦截所有的请求，包括那些指向 HTML、JavaScript、CSS 等资源的请求，但它不会拦截指向 JSP 页面的请求。

5.2.2 ViewResolver

ViewResolver 也称为视图解析器，它的主要职责是解析视图名称，并返回一个 View 对象，这个 View 对象负责将模型数据渲染到客户端。Spring 提供多种 View Resolver 实现，比如：

(1) InternalResourceViewResolver：用于解析传统的 JSP 页面。

(2) BeanNameViewResolver：通过 Bean 的名称来解析视图。

(3) XmlViewResolver：通过 XML 配置来解析视图。

(4) UrlBasedViewResolver：根据 URL 来解析视图。

其中的 InternalResourceViewResolver 为内部资源视图解析器，是日常开发中最常用的视图解析器类型，它是 UrlBasedViewResolver 的子类，通过 prefix 属性指定前缀，suffix 属性指定后缀，然后将前缀 prefix 和后缀 suffix 与具体的视图名称拼接，得到一个视图资源文件的具体加载路径，从而加载真正的视图文件并反馈给用户。示例代码如下：

```
<bean id="viewResolver"
class="org.springframework.web.servlet.view.InternalResourceViewResolver">
      <!--前缀-->
    <property name="prefix" value="/WEB-INF/jsp/"/>
    <!--后缀-->
    <property name="suffix" value=".jsp"/>
 </bean>
```

上面代码设置了视图的前缀和后缀属性，若控制器返回的页面名称为 hello，则经过拼接后的具体路径为/WEB-INF/jsp/hello.jsp。

5.3 控 制 器

在 Spring MVC 中，控制器负责处理客户请求，通常需要定义方法来接收和处理 HTTP 请求，并返回一个模型和视图的响应。控制器是 Spring MVC 架构中的一个关键组件，它是请求处理流程中的入口点。

1. @Controller 注解

可以通过使用@Controller 注解来标记类是否为一个控制器，在使用时只需要把注解加在控制器类上即可，具体格式如下：

```
package   com.controller;
import org.springframework.stereotype.Controller;
@Controller
public class UserController {
    // 处理请求的方法
}
```

Spring MVC 可以使用扫描机制找到应用中所有基于注解的控制器类，需要在配置文件中声明 spring-context，并使用<context:component-scan/>元素指定控制器类的基本包。例如，若控制器类都在 com.controller 包及其子包下，则在 springmvcDemo 应用的配置文件 springmvc-servlet.xml 中添加以下代码：

```
<!-- 使用扫描机制扫描控制器类 -->
<context:component-scan base-package=com.controller" />
```

2. @RequestMapping 注解

为了处理客户请求，使用基于注解的控制器类中可以编写多个处理方法。例如，在以上 UserController 里可实现增加用户、修改用户信息、删除指定用户等方法，每个方法负责不同的请求操作，而@RequestMapping 注解负责映射请求到对应的方法上。该注解有以下几个常见属性：

(1) value 属性：默认属性，用于指定路径，当只有 value 属性时可以省略属性名。例如，通过地址 http://localhost:8080/demo/toUser 访问的请求的处理方法，存在以下两种等价写法：

```
@RequestMapping(value="toUser")
//等价于@RequestMapping("toUser")
public String toUser() {
    return "showUser";
}
```

(2) path 属性：表示指定的路径，该属性支持通配符匹配方式，如@RequestMapping(path="toUser/*")表示 http://localhost:8080/toUser/1 或 http://localhost:8080/toUser/order，两种方式都能够正常访问。

(3) method 属性：表示该方法支持的 HTTP 请求，如 POST、GET、PUT、DELETE 等；若省略 method 属性，则默认支持全部的 HTTP 请求；@RequestMapping(value = "toUser", method = RequestMethod.GET)表示只支持 GET 请求；当注解形式为@RequestMapping(value = "toUser", method = {RequestMethod.GET, RequestMethod.POST})，代表同时支持 GET 和 POST 请求。

(4) params 属性：用于指定请求中的参数。例如，请求中必须包含 type 参数才能执行该请求，即 http://localhost:8080/demo/toUser?type=xxx，其注解代码如下：

```
@RequestMapping(value = "toUser", params = "type")
```

若要求某个参数为固定值，如要求 type 参数为 1 时才能够执行该请求，即 http://localhost:8080/demo/toUser?type=1，则其注解代码如下：

```
@RequestMapping(value = "toUser", params = "type=1")
```

(5) header 属性：表示只有当请求头包含特定的内容时，该请求才会被映射到对应方法。示例注解代码如下：

```
@RequestMapping(value = "/user", headers = "X-Requested-With=XMLHttpRequest")
```

(6) consumers 属性：表示请求的提交内容类型(Content-Type)。例如，下面注解只有当请求的 Content-Type 是 application/json 时，对应方法才会被调用，其注解代码如下：

```
@RequestMapping(value = "toUser", consumes = "application/json")
```

(7) produces 属性：用于指定返回的内容或者编码。示例注解代码如下：

```
@RequestMapping(value = "toUser", produces = "application/json, charset=utf-8")
```

@RequestMapping 可以写在方法上面也可以写在类上面，当写在方法上面时，代表该方法会在收到对应 HTTP 请求后被调用。例如，http://localhost:8080/demo/index/login 请求将调用以下 login 方法：

```
package com.controller;
import org.springframework.stereotype.Controller;
import org.springframework.web.bind.annotation.RequestMapping;
@Controller
public class UserController {
    @RequestMapping(value = "/index/login")
    public String login() {
        return "login";
    }
    @RequestMapping(value = "/index/register")
    public String register() {
        return "register";
    }
}
```

当@RequestMapping 写在类上面时，表示该类中的所有响应请求方法都以该 URL 地址作为基础路径。例如，下面 login 方法对应的请求地址为 http://localhost:8080/demo/index/login：

```
package com.controller;
import org.springframework.stereotype.Controller;
import org.springframework.web.bind.annotation.RequestMapping;
@Controller
@RequestMapping("/index")
public class UserController {
    @RequestMapping("/login")
    public String login() {
        return "login";
    }
    @RequestMapping("/register")
```

```
    public String register() {
        return "register";
    }
}
```

为简化代码的编写，Spring MVC 对@RequestMapping 注解和它的方法属性进行组合，有以下几个组合注解。

- @GetMapping：匹配 GET 请求。
- @PostMapping：匹配 POST 请求。
- @PutMapping：匹配 PUT 请求。
- @DeleteMapping：匹配 DELETE 请求。
- @PatchMapping：匹配 PATCH 请求。

例如，以下注解代码表示只有 POST 请求才能被 add 方法处理：

```
@PostMapping
public void add(@RequestBody Person person) {
}
```

3. 请求处理方法的参数类型

Spring MVC 中，控制器请求处理方法的常见参数类型如下：

(1) ServletRequest/HttpServletRequest 即 Servlet 的请求对象，代表客户端的请求信息。

(2) ServletResponse/HttpServletResponse 即 Servlet 的响应对象，代表向客户端发送的响应信息。示例代码如下：

```
// 处理 HTTP GET 请求，并使用 HttpServletRequest 获取请求参数
@GetMapping("/hello")
public String GetMsg(HttpServletRequest request, HttpServletResponse response) {
    // 从 HttpServletRequest 中获取请求参数
    String clientId = request.getParameter("clientId");
    // 设置响应头
    response.setContentType("text/plain");
    response.setCharacterEncoding("UTF-8");
    // 处理请求并返回响应
    PrintWriter out = response.getWriter();
    out.println("Hello, " + clientId + "!");
    out.flush();
    out.close();
    return null;  //返回一个空的 ModelAndView
}
```

(3) java.util.Map 和 org.springframework.ui.Model：主要用于向视图传递数据。例如，使用 map 对象添加属性“para1”，使用 model 对象添加属性“para2”，相关代码如下：

```
@RequestMapping("/para1")
public String example(Map<String, Object> map) {
    //使用 map 对象添加属性
    map.put("para1", "Hello, World!");
    return "user";  //返回视图名称
}
@RequestMapping("/para2")
public String example(Model model) {
    //使用 model 对象添加属性
    model.addAttribute("para2", "Hello, World!");
    return "user";  //返回视图名称
}
```

(4) HttpSession：表示与请求关联的 HTTP 会话。例如，用于存储与某个客户端会话相关属性的代码如下：

```
@Controller
@RequestMapping("/index")
public class UserController {
    @RequestMapping("/login")
    public String login(HttpSession session, HttpServletRequest request) {
        session.setAttribute("skey", "session 范围的值");
        session.setAttribute("rkey", "request 范围的值");
        return "login";
    }
}
```

(5) @PathVariable：表示获取请求路径中的模板变量，@GetMapping 可以将路径的一部分作为变量。例如，从路径中得到两个变量 schoolId 和 userId，然后通过 @PathVariable 传入 findUser 方法中的代码如下：

```
@GetMapping("/users/{schoolId}/pets/{userId}")
public User findUser(@PathVariable Long schoolId, @PathVariable Long userId) {
    // ...
}
```

(6) @RequestParam：表示获取请求路径中的参数。例如，获取请求路径 http://localhost:8080/demo/users/getuser?userId=100123 中的参数 userId 的代码如下：

```
@GetMapping
    public String setupForm(@RequestParam("userId") int userId, Model model) {
        //...
    }
```

(7) @SessionAttribute：获取会话属性，使用该注解注入会话中的用户信息到 User 对象的示例代码如下：

```
@RequestMapping(value = "/login", method = RequestMethod.POST)
public String login(User user, HttpSession session) {
        //将用户信息存储在会话中
        session.setAttribute("user", user);
        return "redirect:/home";
}
@RequestMapping("/")
public String handle(@SessionAttribute(name = "user")  User user) {
    // ...
}
```

4. 请求处理方法的返回值类型

(1) String 指定要显示的视图、重定向和转发的示例代码如下：

```
@GetMapping("/getusers")
public String listUsers(Model model) {
    //添加模型数据到模型对象
    model.addAttribute("users", userService.getUsers());
    //返回视图名称
    return "users";
}
@GetMapping("/listusers")
public String redirectToUsers() {
    //重定向
    return "redirect:/users/list";
}
@GetMapping("/touser)
public String toUser() {
    //请求转发
    return "forward:/users/user1";
}
```

(2) ModelAndView 可以添加 Model 数据，并指定视图，示例代码如下：

```
@GetMapping("/users")
public ModelAndView listUsers() {
    ModelAndView modelAndView = new ModelAndView();
    modelAndView.addObject("users", userService.getUsers());
    modelAndView.setViewName("users");
    return modelAndView;
}
```

(3) void 表示该方法没有直接返回一个模型或者视图，而是通过其他方式来处理响应，例如重定向、写入响应内容等，示例代码如下：

```
@RequestMapping(value = "/redirect", method = RequestMethod.GET)
public void redirectToHome( HttpServletResponse response) throws IOException {
//重定向
    response.sendRedirect("/home");
}
@RequestMapping(value = "/writeResponse", method = RequestMethod.GET)
 public void writeResponse(HttpServletResponse response) throws IOException {
//直接通过输出流写入响应内容
        response.setContentType("text/plain");
        response.getOutputStream().write("Hello, World!".getBytes());
}
```

5.4 接收请求数据

在 Spring MVC 中，控制器方法可以通过多种方式获取 HTTP 请求的查询数据或者表单数据，可以通过不同数据类型的形参将请求消息数据与后台方法参数建立连接，从而达到接收请求数据的目的，通常有 3 种方法：HttpServletRequest 方式、绑定简单数据类型方式和绑定实体类对象方式。

5.4.1 HttpServletRequest 方式

HttpServletRequest 接口是 Java Servlet 规范中定义的一个接口，用于代表客户端的请求。在 Java Web 应用程序中，开发者可以使用这个接口来获取客户端发送请求时所包含的信息，如请求参数。可以直接通过 HttpServletRequest 的 getParameter 方法获取单个参数的值。可以通过 getParameterMap 方法获取所有表单参数的 Map<String, String[]>，其中关键字是参数名称；值是一个字符串数组，对应于该参数的所有值(适用于多选框等)。

例如，下面定义的控制器参数，在方法内部利用 HttpServletRequest 对象获取客户端发送的请求参数，具体代码如下：

```
@RequestMapping("/selectName")
public String getGender(HttpServletRequest request, HttpServletResponse response,
Model model) {
        System.out.println("请求参数"+request.getParameter("name"));
        model.addAttribute("msg", request.getParameter("name"));
        return "index";
 }
@RequestMapping("/selectUser")
public String selectUser (HttpServletRequest request, HttpServletResponse response,
Model model) {
```

```
        // 使用 getParameterMap() 方法，可以获取所有参数
        Map<String, String[]> parameterMap = request.getParameterMap();
        //遍历 parameterMap 来获取所有参数的值
        for (Map.Entry<String, String[]> entry : parameterMap.entrySet()) {
            String name = entry.getKey();
            String[] values = entry.getValue();
            // 打印参数名和所有对应的值
            for (String value : values) {
                System.out.println(name + ": " + value);
            }
        model.addAttribute("msg", "hello");
        return "welcome";
    }
```

以上代码中，request.getParameter("name")获取了请求中名为"name"的参数值，并将其打印出来；然后这个值被添加到模型 model 中，指定键为"msg"，这样就可以在视图中访问这个值。request.getParameterMap()用来获取所有请求参数的 Map，通过遍历这个 Map，可以打印出每个参数的名称和所有对应的值。

5.4.2　绑定简单数据类型方式

1. 基本数据类型

对于请求中的 GET 和 POST 方法，Spring MVC 可以自动将参数值绑定到简单数据类型(如 int、double、boolean 等)和 String 类型的控制器方法参数上。例如，下面是控制器自动绑定 Integer 类型 id 参数的代码：

```
@RequestMapping("/getUser")
publicString getUser (Integer id) {
    System.out.println("id="+id);
    return "success";
}
```

若前端请求中的参数名和后台控制器类方法中的形参名不一样，如请求地址为 http://localhost:8080/demo/getUser?user_id=1，则需要在控制器方法参数前面添加 Spring MVC 提供的@RequestParam 注解类型，具体代码如下：

```
@RequestMapping("getUser")
publicString getUser(@RequestParam(value="user_id")Integer id) {
    System.out.println("id="+id);
    return "success";
}
```

2. 数组类型

在实际开发时，可能会遇到前端请求需要传递到后台一个或多个相同名称参

数的情况。例如，包含以下兴趣爱好的前端页面的代码如下：

```
<form action="${pageContext.request.contextPath }/addHobby"method="post">
  <table width="20%"border=1>
    <tr><td>选择</td><td>兴趣爱好</td></tr>
    <tr>
      <td><input name="hobbies"   value="1 " type="checkbox"></td>
      <td>篮球</td>
    </tr>
    <tr>
      <td><input name=" hobbies " value="2"   type="checkbox">
      </td><td>足球</td>
    </tr>
    <tr>
      <td><input name=" hobbies " value="3" type="checkbox"></td>
      <td>羽毛球</td>
    </tr>
  </table>
  <inputtype="submit" value="添加"/>
</form>
```

针对上述情况，可以使用数组作为控制器方法参数，后台就可以进行绑定接收，具体如下：

```
@RequestMapping("/addHobby")
public String addHobby(Integer[]ids) {
  if(ids !=null){
      for (Integer id : ids) {
        System.out.println("添加了 id 为"+id+"的兴趣！ ");}
    }
    else{
      System.out.println("ids=null");
    }
    return "success";
}
```

5.4.3 绑定实体类对象方式

1. 基本 POJO

如果一个请求有多个不同类型，则绑定简单数据类型参数时显得比较烦琐。例如，有 4 个表单参数，按照绑定简单数据类型方式，控制器方法中会出现 userId、userName、age 和 password 等 4 个参数，具体代码如下：

```
<form th:action="@{/getFormalParam}" method="post">
    <table style="margin: auto">
        <tr>
            <td>用户 ID：</td>
            <td><input type="text" name="userId" required><br></td>
        </tr>
        <tr>
            <td>用户名：</td>
            <td><input type="text" name="userName" required><br></td>
        </tr>
        <tr>
            <td>年龄：</td>
            <td><input type="text" name="age" required><br></td>
        </tr>
        <tr>
            <td>密码：</td>
            <td><input type="password" name="password" required><br></td>
        </tr>
        <tr>
            <td colspan="2" align="center"><input type="submit"></td>
        </tr>
</table>
```

可使用 POJO(Plain Old Java Object)提供一种结构化的方式来封装和传递请求数据，将所有关联的请求参数封装在一个 POJO 中，具体实现方法如下：

(1) 定义 POJO 类，保证请求参数的名称与实体类中的属性名一致，并配置 setter 和 getter 方法，具体代码如下：

```
public class User {
    private String UserId;
    private String UserName;
    private Integer age;
    private String password;
    public String getUserId() {
        return UserId;
    }
    public void setUserId(String userId) {
        UserId = userId;
    }
    public String getUserName() {
        return UserName;
    }
```

```
    public void setUserName(String userName) {
        UserName = userName;
    }
    public Integer getAge() {
        return age;
    }
    public void setAge(Integer age) {
        this.age = age;
    }
    public Integer getPassword () {
        return password;
    }
    public void setPassword (String password) {
        this. password = password;
    }
}
```

(2) 在方法中直接使用该 POJO 作为形参来完成接收请求数据，具体代码如下：

```
@RequestMapping("/getUser")
public String getUser(User user) {
    System.out.println("userId： " + user.getUserId());
    System.out.println("userName： " + user.getUserName());
    System.out.println("password： " + user.getPassword());
    return "success";
}
```

注意事项：请求参数名和 POJO 的属性名要严格保持一致，才能自动将请求参数绑定到 POJO 中，否则后台接收到的参数为 null。

2. 包装 POJO

在实际使用时，可能需要将一个 POJO 作为另一个 POJO 的属性，即包装 POJO。例如，下面实体类代码在用户对象中包含地址对象，这样在使用时，就可以通过用户查询到地址信息，具体代码如下：

```
public class User {
    private Integer userID;
    private String username;
    private Integer age;
    private Address address;  //个人关联的地址
    //...省略 getter/setter 方法
}
//地址类：
public class Address {
    private String province;
```

```
    private String city;
    //...省略 getter/setter 方法
}
```

对于对象的直接属性，在前端代码中参数名称要与属性名严格一致；对于子对象的属性，参数名必须是“对象.属性”的形式，具体代码如下：

```
<form action="${pageContext.request.contextPath }/registerUser"method="post">
身份证号：<input type="text" name="userID"><br>
用户名：<input type="text" name="username"><br>
        年龄：<input type="text" name="age"><br>
        省份：<input type="text" name="address.province"><br>
        城市：<input type="text" name="address.city"><br>
        <input type="submit" value="提交">
</form>
```

对于以上实体类和前端页面，对应控制器类方法格式如下：

```
@RequestMapping("/registerUser")
publicString registerUser (User user) {
   System.out.println("用户名="+user.getUsername());
   System.out.println("年龄="+user.getAge());
   System.out.println("省份="+user.getAddress().getProvince());
   System.out.println("城市="+user.getAddress().getCity());
   return"success";
}
```

3. 包装集合

由于集合类型不能直接进行参数绑定，因此需要创建一个包含 List 或其他集合类型属性的类，例如 UserList 类，其中包含一个 List 类型的属性 users，并为其提供 getter 和 setter 方法：

```
public class User {
    private String userName;
    private int userAge;
    // getter 和 setter 方法省略
}
public class UserList {
    private List<User> users;
// getter 和 setter 方法省略
}
```

编写表单时，需要确保表单中的 name 属性值与绑定的模型类属性名一致，具体代码如下：

```
<form action="${pageContext.request.contextPath }/editUsers"method="post">
    <input type="text" name="users[0].userName" />
```

```
    <input type="text" name="users[0].userAge" />
    <input type="text" name="users[1].userName" />
    <input type="text" name="users[1].userAge" />
    <input type="submit" value="Submit" />
</form>
```

需要在 Controller 类方法中绑定 UserList 类的实例，Spring MVC 会自动将请求中的参数绑定到 UserList 类的 users 属性上，具体代码如下：

```
@Controller
public class UserController {
    @PostMapping("/editUsers")
    public String list(UserList userList, Model model) {
        //获取封装在 POJO 中的集合
        List<User> users=userList.getUsers();
        model.addAttribute("userList", users);
        return "result";
    }
}
```

当接收到请求时，Spring MVC 会根据配置的绑定机制，将请求参数解析并填充到 UserList 类的 users 属性中，这样就可以在控制器方法中直接使用这个集合类型的参数。

5.4.4 接收请求数据综合案例

下面通过一个案例帮助读者掌握接收请求数据的编程方法，该案例模拟实现购物网站若干功能，若在网站首页面输入关键词进行搜索，则 http 请求将绑定关键词，服务器端处理后跳转到推荐商品页面；若在首页面选择类别，则 http 请求将绑定类别数组，服务器端处理后跳转到推荐商品页面；代码中还要完成添加购物车、购物车管理、订单管理等功能，具体实现步骤如下：

(1) 参照 5.1.2 节的步骤创建名称为 WebDataBind 的 Maven 项目并添加依赖。

(2) 在 pom.xml 中增加 jstl 依赖。

pom.xml 修改

```
<dependency>
        <groupId>javax.servlet</groupId>
        <artifactId>jstl</artifactId>
        <version>1.2</version>
</dependency>
```

以上配置声明了 JSTL(JSP Standard Tag Library)的依赖，JSTL 是一个用于 Java 服务器端开发的标准化标签库，它提供了一系列标签用于简化 JSP 页面中的常见任务，如条件判断、遍历、格式化等。

(3) 在项目的 java 文件夹中创建一个包 com.DataBind.pojo，在该包下创建实

体类 Product.java、AddressInfo.java、Order.java 和 Cart.java，分别代表产品类、地址信息类、订单类、购物车类。

Product.java

```
package com.DataBind.pojo;
public class Product {
    private int id;
    private String name;
    private double price;
    private  int count;
    public Product() {
    }
    public Product(int id, String name, double price, int count) {
        this.id = id;
        this.name = name;
        this.price = price;
        this.count = count;
    }
    public int getId() {
        return id;
    }
    public void setId(int id) {
        this.id = id;
    }
    public String getName() {
        return name;
    }
    public void setName(String name) {
        this.name = name;
    }
    public double getPrice() {
        return price;
    }
    public void setPrice(double price) {
        this.price = price;
    }
    public int getCount() {
        return count;
    }
    public void setCount(int count) {
```

```
        this.count = count;
    }
}
```

AddressInfo.java

```
package com.DataBind.pojo;
public class AddressInfo {
    String address;
    String phone;
    public String getAddress() {
        return address;
    }
    public void setAddress(String address) {
        this.address = address;
    }
    public String getPhone() {
        return phone;
    }
    public void setPhone(String phone) {
        this.phone = phone;
    }
}
```

Order.java

```
package com.DataBind.pojo;
public class Order {
    AddressInfo addressinfo;
    String description;
    String orderNumber;
    double total;
    public double getTotal() {
        return total;
    }
    public void setTotal(double total) {
        this.total = total;
    }
    public String getOrderNumber() {
        return orderNumber;
    }
    public void setOrderNumber(String orderNumber) {
        this.orderNumber = orderNumber;
```

```
    }
    public AddressInfo getAddressinfo() {
        return addressinfo;
    }
    public void setAddressinfo(AddressInfo addressinfo) {
        this.addressinfo = addressinfo;
    }
    public String getDescription() {
        return description;
    }
    public void setDescription(String description) {
        this.description = description;
    }
}
```

Cart.java

```
package com.DataBind.pojo;
import java.util.List;
public class Cart {
    List<Product> products;
    public List<Product> getProducts() {
        return products;
    }
    public void setProducts(List<Product> products) {
        this.products = products;
    }
}
```

(4) 在已创建的项目 java 文件夹中创建一个包 com.DataBind. controller，在该包下创建名称为 MyController 的控制器类。

MyController.java

```
package com.DataBind.controller;
import com.DataBind.pojo.Cart;
import com.DataBind.pojo.Order;
import com.DataBind.pojo.Product;
import org.springframework.stereotype.Controller;
import org.springframework.ui.Model;
import org.springframework.web.bind.annotation.ModelAttribute;
import org.springframework.web.bind.annotation.RequestMapping;
import org.springframework.web.bind.annotation.RequestMethod;
import org.springframework.web.bind.annotation.RequestParam;
```

```
import java.util.ArrayList;
import java.util.List;
@Controller
public class MyController {
    public static    List<Product> products=new ArrayList<Product>();
    public static List<Product> productList=new ArrayList<Product>();
    // 1. 当用户选择推荐商品类别时传递数组到后台
    @RequestMapping(value = "/recommend", method = RequestMethod.POST)
    public String recommend(@RequestParam("hobbies") int[] hobbies, Model model) {
        System.out.println("选择了以下类别：");
        String msg="";
        for(int hobby : hobbies ){
            msg=msg+hobby+" ";
            for(int i=0; i<3; i++){
                Product product=new Product(hobby*1000+i, "商品"+(char)('A'+hobby)+
                i, Math.round(10000*Math.random())/100, 0);
                productList.add(product);
            }
        }
        System.out.println(msg);
        model.addAttribute("productList", productList);
        model.addAttribute("msg", msg);
        return "home";
    }
    // 2. 当用户搜索感兴趣商品时传递简单数据到后台
    @RequestMapping(value = "/search", method = RequestMethod.GET)
    public String search(@RequestParam("keyword") String keyword, Model model) {
        String msg=keyword+" 有关";
            // 处理请求，根据关键词查询商品并返回结果
        System.out.println("页面将按照搜索结果"+keyword+"列出商品");
        for(int i=0; i<3; i++){
            Product product=new Product(1000+i, "商品 D"+i, 100*Math.random(), 0);
            productList.add(product);
        }
        model.addAttribute("productList", productList);
        model.addAttribute("msg", msg);
        return "home";
    }
    // 3. 当用户单击加入购物车功能时传递 POJO 到后台
```

```
@RequestMapping(value = "/addToCart", method = RequestMethod.POST)
public String addToCart(@ModelAttribute("product") Product product, Model model) {
    // 处理请求，将商品添加到购物车并返回结果
    System.out.println("购物车大小="+products.size()+" 添加商品"+product.getName());
    int flag=0;
    for(Product old: products){
        if (old.getId()==product.getId()){
            flag=1;
            old.setCount(old.getCount()+1);
            System.out.println(old.getId()+" +1");
            break;
        }
    }
    if(flag==0) {
        product.setCount(1);
        products.add(product);
    }
    model.addAttribute("productList", productList);
    return "home";
}
// 4. 当用户单击购物车管理时，获取商品列表并返回购物车页面
@RequestMapping(value = "/toCart", method = RequestMethod.GET)
public String toCart(Model model) {
    // 处理请求，获取购物车中的商品列表并返回结果
    System.out.println("购物车共有商品"+products.size()+"种");
    model.addAttribute("products", products);
    return "cart";
}
// 5. 当用户生成订单时传递包装列表的 POJO 到后台
@RequestMapping(value = "/manageOrder", method = RequestMethod.POST)
public String manageCart(Cart cart, Model model) {
    products= cart.getProducts();
    double total=0;
    for(Product product: products){
        total=total+product.getPrice()+product.getPrice();
    }
    String description="选购了"+products.get(0).getName()+"等"+products.size()+"件商品";
    model.addAttribute("description", description);
    String orderNumber="order00001";
```

```
        model.addAttribute("orderNumber", orderNumber);
        model.addAttribute("total", total);
        return "order";
    }
    // 6. 当提交订单时传递包装 POJO 到后台
    @RequestMapping(value = "/submitOrder", method = RequestMethod.POST)
    public String submitOrder(Order order, Model model) {
        // 处理请求，提交订单并返回结果
        String address=order.getAddressinfo().getAddress();
        double total=order.getTotal();
        model.addAttribute("total", total);
        model.addAttribute("address", address);
        return "pay";
    }
}
```

以上控制器类代码，当用户选择推荐商品类别时传递数组到后台，当用户搜索感兴趣的商品时传递简单数据到后台；后台经过控制器返回到主页面，并使用 model 传递列表 productList 到页面。当用户单击加入购物车功能时传递 POJO 到后台，后台将购物车数据统计计数并存储到 POJO 列表(即商品列表 products)中。当用户单击购物车管理时后台将返回购物车管理页面，并通过 model 传递存储的商品列表 products。当用户单击生成订单时绑定包装列表的 POJO(即 cart 对象)到后台，后台生成订单数据并返回订单页面。当提交订单时传递包装地址信息的 POJO(即 order 对象)到后台，后台接收数据，返回付款页面。

(5) 参照 5.1.2 节，创建 Spring 的配置文件 spring-config.xml。

(6) 参照 5.1.2 节，添加、修改 web 目录和 web.xml 位置，并添加 web.xml 文件用于配置前端控制器、编码过滤器和启动页面。

web.xml

```
<?xml version="1.0" encoding="UTF-8"?>
<web-app xmlns="http://xmlns.jcp.org/xml/ns/javaee"
         xmlns:xsi="http://www.w3.org/2001/XMLSchema-instance"
         xsi:schemaLocation="http://xmlns.jcp.org/xml/ns/javaee
         http://xmlns.jcp.org/xml/ns/javaee/web-app_4_0.xsd"
         version="4.0">
    <display-name>demo2DataBind</display-name>
    <servlet>
        <!-- 前端控制器 -->
        <servlet-name>springmvc</servlet-name>
        <servlet-class>org.springframework.web.servlet.DispatcherServlet</servlet-class>
        <!-- 前端控制器配置文件 -->
```

```
        <init-param>
            <param-name>contextConfigLocation</param-name>
            <param-value>classpath:applicationContext.xml</param-value>
        </init-param>
        <!-- 容器启动时立即加载前端控制器 -->
        <load-on-startup>1</load-on-startup>
    </servlet>
    <!-- 前端控制器对所有 URL 拦截，除了 JSP 之外-->
    <servlet-mapping>
        <servlet-name>springmvc</servlet-name>
        <url-pattern>/</url-pattern>
    </servlet-mapping>
    <!-- 编码过滤器 -->
    <filter>
        <filter-name>myfilter</filter-name>
        <filter-class>org.springframework.web.filter.CharacterEncodingFilter</filter-class>
        <init-param>
            <param-name>encoding</param-name>
            <param-value>UTF-8</param-value>
        </init-param>
    </filter>
    <filter-mapping>
        <filter-name>myfilter</filter-name>
        <url-pattern>/*</url-pattern>
    </filter-mapping>
    <welcome-file-list>
        <welcome-file>welcome.jsp</welcome-file>
    </welcome-file-list>
</web-app>
```

(7) 在 web 目录下创建欢迎页面，对应 welcome.jsp 文件。

welcome.jsp

```
<%@ page language="java" contentType="text/html; charset=UTF-8" pageEncoding="UTF-8"%>
<!DOCTYPE html>
<html>
<head>
    <meta charset="UTF-8">
    <title>商品列表</title>
</head>
<body>
```

```
<h1>商品列表</h1>
<form action="${pageContext.request.contextPath}/search" method="get">
    <label for="keyword">搜索关键词：</label>
    <input type="text" id="keyword" name="keyword">
    <input type="submit" value="搜索">
</form>
<form action="${pageContext.request.contextPath}/recommend" method="post">
    <p>选择感兴趣商品类别：</p>
    <input type="checkbox" id="categoryA" name="hobbies" value="1">
    <label for="categoryA">类别 A</label>
    <input type="checkbox" id="categoryB" name="hobbies" value="2">
    <label for="categoryB">类别 B</label>
    <input type="checkbox" id="categoryC" name="hobbies" value="3">
    <label for="categoryC">类别 C</label>
    <input type="submit" value="提交">
</form>
</body>
</html>
```

以上代码包含两个表单：搜索表单和推荐表单。搜索表单允许用户输入关键词来搜索商品，当用户输入关键词并单击提交按钮时，会将数据发送到服务器的 /search 路径；推荐表单允许用户选择感兴趣的商品类别，当用户单击提交按钮时，会将数据发送到服务器的 /recommend 路径。

(8) 在 /WEB-INF 文件夹下创建名称为 jsp 的子文件夹，在这个子文件夹下创建 home、cart、order 和 pay 4 个视图文件，分别为主页面、购物车页面、订单页面和付款页面。

home.jsp

```
<%@ page contentType="text/html;charset=UTF-8" language="java" %>
<%@ taglib prefix="c" uri="http://java.sun.com/jsp/jstl/core" %>
<html>
<head>
    <title>Title</title>
</head>
<body>
<h1>${msg} 类型的商品列表</h1>
<table border="1">
    <tr>
        <th>商品 ID</th>
        <th>商品名称</th>
        <th>价格</th>
```

```
            <th>操作</th>
        </tr>
         <%-- 这个地方用到了 JSTL 和 EL 表达式--%>
        <c:forEach var="product" items="${productList}">
            <form action="${pageContext.request.contextPath}/addToCart" method="post">
                <tr>
                    <td><input type="text" value="${product.id}" name="id" ></td>
                    <td><input type="text" value="${product.name}" name="name" ></td>
                    <td><input type="text" value="${product.price}" name="price" ></td>
                    <input type="hidden" value="${product.count}" name="count" >
                    <td><input type="submit"   value="加入购物车"></td>
                </tr>
            </form>
        </c:forEach>
    </table>
    <a href="${pageContext.request.contextPath}/toCart" >购物车管理</a>
    </body>
    </html>
```

以上 home 页面提供了一个商品列表的功能，允许用户查看商品的详细信息，并将商品添加到购物车中。

cart.jsp

```
<%@ page contentType="text/html;charset=UTF-8" language="java" %>
<%@ taglib prefix="c" uri="http://java.sun.com/jsp/jstl/core" %>
<html>
<head>
    <title>购物车管理</title>
    <!-- 必要的 CSS 和 JavaScript -->
</head>
<body>
<h1>我的购物车</h1>
<div>
    <form action="${pageContext.request.contextPath}/manageOrder" method="post">
        <!-- 显示购物车中的订单列表 -->
        <table>
            <tr>
                <th>商品 id</th>
                <th>商品名</th>
                <th>价格</th>
                <th>数量</th>
```

```
                <th>增加</th>
                <th>减少</th>
            </tr>
            <%-- 这个地方用到了 JSTL 和 EL 表达式--%>
            <c:forEach var="product" items="${products}" varStatus="status">
                <tr>
                    <td><input type="checkbox" value="${product.id}" name=
                    "products[${status.count-1}].id"
                                checked="checked" id="${product.id}">
                        <label for="${product.id}">${product.id}</label>
                    </td>
                    <td><input type="text" value="${product.name}" name=
                    "products[${status.count-1}].name"
                                readonly="true"></td>
                    <td><input type="text" value="${product.price}" name=
                    "products[${status.count-1}].price"
                                readonly="true"></td>
                    <td><input type="text" value="${product.count}" name=
                    "products[${status.count-1}].count"
                                id="no-${status.count}" min="1"></td>
                    <td>
                        <button type="button" onclick=
                        "myclick(1, 'no-${status.count}')">+</button>
                    </td>
                    <td>
                        <button type="button" onclick=
                        "myclick(2, 'no-${status.count}')">-</button>
                    </td>
                </tr>
            </c:forEach>
        </table>
        <input type="submit" value="生成订单">
    </form>
</div>
<script>
    function myclick(select, id) {
        var countDom = document.getElementById(id)
```

```
            if (select == 1) {
                countDom.value = parseInt(countDom.value) + 1
            } else if (select == 2) {
                countDom.value = parseInt(countDom.value) - 1
            }
        }
    </script>
    </body>
    </html>
```

以上 cart 页面对应一个购物车管理的功能，允许用户查看购物车中的商品列表，修改商品数量，并生成订单。

order.jsp

```
<%@ page language="java" contentType="text/html; charset=UTF-8"pageEncoding="UTF-8"%>
<!DOCTYPE html>
<html>
<head>
<meta charset="UTF-8">
    <title>我的订单</title>
</head>
<body>
<h2>订单详情</h2>
<form action="${pageContext.request.contextPath}/submitOrder" method="post">
    <label for="address">地址：</label><br>
    <input type="text" id="address" name="addressinfo.address" required><br><br>
    <label for="telephone">电话：</label><br>
    <input type="tel" id="telephone" name="addressinfo.phone" required><br><br>
    <label for="orderNumber">订单号：</label><br>
    <input type="text" id="orderNumber" name="orderNumber" readonly value=
    "${orderNumber}"><br><br>
    <label for="description">描述：</label><br>
    <input type="text" id="description" name="description" readonly value=
    "${description}"><br><br>
    <label for="total">总金额：</label><br>
    <input type="text" id="total" name="total" readonly value="${total}"><br><br>
    <input type="submit" value="提交订单">
</form>
</body>
```

```
</html>
```

以上 order 页面功能是让用户查看订单的摘要信息，并提供输入表单以便用户输入必要的收货信息。用户输入信息后，可以通过单击提交按钮将订单和收货信息一起提交到服务器进行处理。

pay.jsp

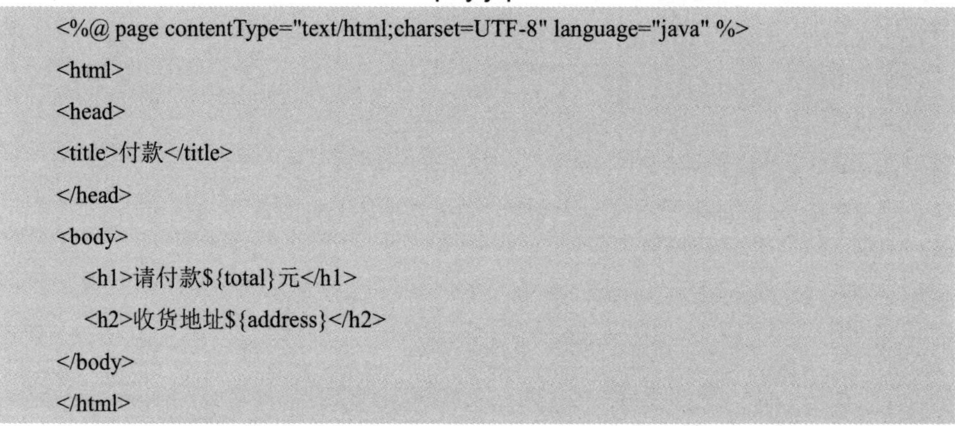

```
<%@ page contentType="text/html;charset=UTF-8" language="java" %>
<html>
<head>
<title>付款</title>
</head>
<body>
  <h1>请付款${total}元</h1>
  <h2>收货地址${address}</h2>
</body>
</html>
```

以上 pay 页面使用表达式 ${total}动态显示付款金额，使用 EL 表达式 ${address}动态显示收货地址。

(9) 参照 5.1.2 节配置 artifacts 和 tomcat，最终项目结构如图 5-12 所示。

(10) 运行代码后出现如图 5-13 所示的欢迎页面，即 welcome.jsp 页面。

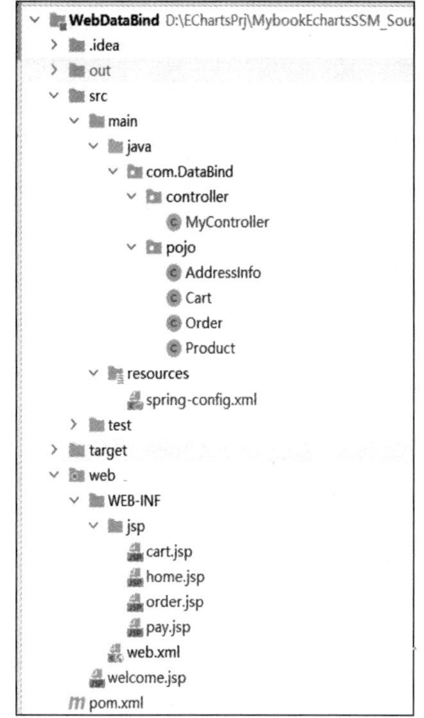

图 5-12 项目结构

图 5-13 欢迎页面

若在首页面输入关键词后单击搜索按钮，http 请求将绑定简单数据，服务器处理后跳转到如图 5-14 所示的 home.jsp 页面。

图 5-14　搜索后的首页面

若在首页面选择类别，则 http 请求中将绑定数组，服务器处理后跳转到如图 5-15 所示的 home.jsp 页面。

图 5-15　选择类别后的首页面

单击若干加入购物车按钮，http 请求将绑定 POJO 类型数据，选择购物车管理选项，出现如图 5-16 所示的购物车页面。

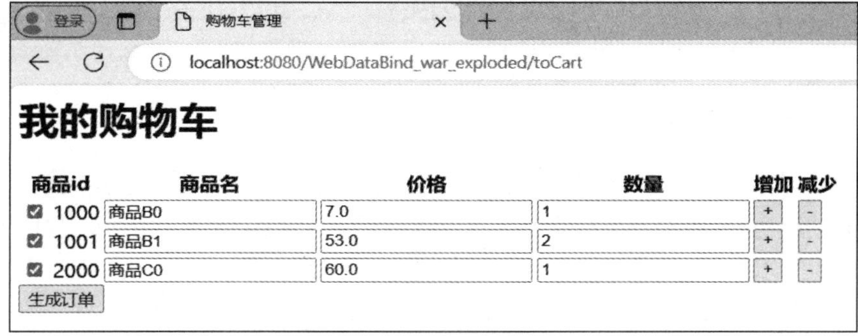

图 5-16　购物车页面

单击生成订单按钮，http 请求将绑定包装 POJO 列表的数据，服务器处理后出现如图 5-17 所示的订单页面。

图 5-17　订单页面

输入订单地址后单击提交订单按钮，http 请求将绑定包装 POJO 的数据，服务器处理后将出现如图 5-18 所示的付款页面。

图 5-18　付款页面

5.5　JSON 数据转换和 RESTful 实现

5.5.1　JSON 数据交互

JSON 全称为 JavaScript Object Notation，是一种轻量级的数据交换格式，它易于开发人员阅读和编写，同时也易于机器解析和生成。JSON 有数组和对象两种结构：数组在 JSON 中是有序的集合，值之间用逗号分隔，整个数组由中括号包裹；对象在 JSON 中由一组无序的键值对构成，每个键值对由冒号分隔，键和值都用双引号标记。示例结构如下：

```
//JSON 数组
[1, "two", 3.0, true, null]
//JSON 对象
{"name": "John", "age": 30, "city": "New York"}
```

JSON 格式的数据在 Web 服务和 Ajax 请求中广泛使用，可以表示结构化的数据。为了实现浏览器与控制器类之间的 JSON 数据交互，Spring MVC 提供了一个 MappingJackson2HttpMessageConverter 类，既可以将 Java 对象转换为 JSON 数据，也可以将 JSON 数据转换为 Java 对象，使得 Web 应用程序能够正确地处理 JSON 格式的请求和响应。该类的使用依赖于 jackson 开源包，具体如下。

(1) jackson-annotations-x.x.x.jar：注解包。

(2) jackson-core-x.x.x.jar：核心包。

(3) jackson-databind-x.x.x.jar：数据绑定包。

另外，Spring MVC 提供了与格式转换相关的两个注解，分别是@Request Body 和@ResponseBody，具体说明如下。

(1) @RequestBody：用于将请求体中的数据绑定到控制器方法的形参上，即完成将请求内容转为 Java 对象。例如，当客户端向 /api/users 发送一个 POST 请求时，@RequestBody 注解指示 Spring 框架将请求体中的 JSON 数据转换为 User 对象，相关代码如下：

```
@PostMapping("/users")
public User createUser(@RequestBody User user) {
    // 处理用户信息
    // 返回 user 对象
    return user;
}
```

(2) @ResponseBody：写在方法上，用于将控制器方法的返回值作为响应报文的响应体，即将返回对象转化为合适的格式。例如，当客户端向 /api/users/{id} 发送一个 GET 请求时，方法返回的 User 对象会被 Spring 框架自动转换为 JSON 格式的字符串，并直接作为 HTTP 响应体返回给客户端，相关代码如下：

```
@GetMapping("/users/{id}")
@ResponseBody
public User getUser(@PathVariable("id") Long id) {
    // 查询用户信息
    User user = queryUser(id);
    return user;
}
```

多学一招：Content-Type 一般是指网页中存在的内容类型，用于定义网络文件的类型和网页的编码，通常有以下几种常见形式：

- text/html 为 HTML 文档，是网页最常用的 MIME 类型。
- text/plain 为纯文本数据，没有格式化或编码。

• application/json为JSON格式的数据，是一种轻量级的数据交换格式。

• application/xml为XML格式的数据，是一种用于网络传输和数据存储的标记语言。

• application/pdf为PDF文件，是一种用于文档交换的文件格式。

• image/jpeg和image/png分别为JPEG和PNG图像文件。

• audio/mpeg为MP3音频文件，即MPEG-1 Audio Layer 3的文件。

• video/mp4为MP4视频文件，是一种广泛使用的视频文件格式。

• application/octet-stream为二进制数据，表示传输的数据是原始的二进制流，例如软件下载。

• application/x-www-form-urlencoded为HTML表单提交的数据，数据被编码为键值对的形式。

• multipart/form-data为包含二进制数据的表单，通常用于在表单数据中包含上传文件的场景。

【案例5-1】 下面通过一个用户注册的案例，完成前端页面和服务端JSON类型信息的传递，具体实现步骤如下：

(1) 参照5.1.2节创建一个名称为WebJson的Maven项目。

(2) 在5.1.2节的pom.xml依赖基础上增加关于JSON的依赖。具体代码如下：

```
<dependency>
    <groupId>com.fasterxml.jackson.core</groupId>
    <artifactId>jackson-annotations</artifactId>
    <version>2.9.2</version>
</dependency>
<dependency>
    <groupId>com.fasterxml.jackson.core</groupId>
    <artifactId>jackson-core</artifactId>
    <version>2.9.2</version>
</dependency>
<dependency>
    <groupId>com.fasterxml.jackson.core</groupId>
    <artifactId>jackson-databind</artifactId>
    <version>2.9.2</version>
</dependency>
```

以上代码中的jackson-annotations依赖包含了Jackson的注解功能，它提供了一组注解，可以用于自定义Java对象与JSON之间的映射关系；而jackson-core依赖是Jackson的核心库，它提供了处理JSON数据的底层功能，包括JSON解析和生成；jackson-databind依赖是Jackson的数据绑定库，它建立在jackson-core和jackson-annotations之上，提供了高级的映射功能，它允许开发者轻松地将JSON数据绑定到Java对象(POJO)上。

(3) 在web目录下新建js目录，导入项目需要的jquery.jar包。

(4) 在 WEB-INF 目录下新建 jsp 目录，然后创建 JSON 测试页面 index.jsp。

index.jsp

```
<%@ page language="java" contentType="text/html; charset=UTF-8" pageEncoding="UTF-8"%>
<!DOCTYPE html>
<html>
<head>
    <meta charset="UTF-8">
    <title>Insert title here</title>
    <script type="text/javascript" src="${pageContext.request.contextPath}/js/jquery-3.3.1.js">
    </script>
    <script type="text/javascript">
        function    testJson(){
            var username=$("#username").val();
            var password=$("#password").val();
            var phone=$("#phone").val();
            $("#userid").text("hello");
            $.ajax({
                url:"${pageContext.request.contextPath}/userJson",
                type:"post",
                //发送的数据
               data:JSON.stringify({username:username, password:password, phone:phone}),
                //发送的格式
                contentType:"application/json; charsert=UTF-8",
                //回调响应格式
                dataType:"json",
                //回调函数
                success:function(res){
                    if(res!=null){
                        $("#reslut").html("尊敬的"+res.username+", 您已注册成功, 用户 ID 为:
                        "+res.idString);
                    }
                }
            });
        }
    </script>
</head>
<body>
<h1>注册</h1>
<form>
    用户名: <input type = "text" name = "username" id = "username"/><br/>
    密       码: <input type = "password" name = "password" id =
"password" /> <br/>
```

```
    电      话： <input type = "text" name = "phone" id = "phone"/> <br/>
</form>
<button onclick = "testJson()" type = "button">提交</button>
<p id = "reslut"></p>
</body>
</html>
```

以上代码提供了一个用户注册的功能，并且使用 Ajax 技术实现向服务器发送和接收 JSON 数据。在<form>中按钮的默认事件里会执行 form.sumbit()，这就会导致表单的提交，同时伴随页面的刷新使 Ajax 的结果被覆盖，因此这个地方 button 的 type 属性用“button”。

(5) 创建 demoJson.pojo 包，然后创建实体类 User.java。

User.java

```
package demoJson.pojo;
public class User {
    public String username;
    public String password;
    public String idString;
    public String phone;
    public String getPhone() {
        return phone;
    }
    public void setPhone(String phone) {
        this.phone = phone;
    }
    public String getIdString() {
        return idString;
    }
    public void setIdString(String idString) {
        this.idString = idString;
    }
    public String getUsername() {
        return username;
    }
    public void setUsername(String username) {
        this.username = username;
    }
    public String getPassword() {
        return password;
    }
```

```
    public void setPassword(String password) {
        this.password = password;
    }
    @Override
    public String toString() {
        return "User [username=" + username + ", phone=" + phone + ", idString=" + idString + "]";
    }
}
```

(6) 创建 demoJson.control 包，然后创建名为 MyController 的控制器类 MyController.java。

MyController.java

```
package demoJson.control;
import demoJson.pojo.User;
import org.springframework.stereotype.Controller;
import org.springframework.web.bind.annotation.RequestBody;
import org.springframework.web.bind.annotation.RequestMapping;
import org.springframework.web.bind.annotation.ResponseBody;
import org.springframework.web.servlet.ModelAndView;
import java.util.Date;
@Controller
public class MyController {
    @RequestMapping("/")
    public ModelAndView testJson() throws Exception{
        ModelAndView modelAndView=new ModelAndView();
        modelAndView.addObject("msg", "");
        modelAndView.setViewName("index");
        return modelAndView;
    }
    @RequestMapping("/userJson")
    @ResponseBody
    public User confirmUser(@RequestBody User user) {
        user.setIdString(String.valueOf(new Date().getTime()));
        System.out.println(user.toString());
        return user;
    }
}
```

以上控制器类控制器提供了两个路径的处理，根路径“/”的请求对应 testJson()方法，它返回一个名为"index"的视图；另一个路径“/userJson”的请求对应 confirmUser()方法，它接收一个 JSON 格式的 User 对象，处理后返回一个 JSON 格式的 User 对象响应。

(7) 修改 Spring 配置文件 applicationContext.xml：添加关于组件扫描、注解驱动、静态资源处理和视图解析器等配置。

applicationContext.xml

```
<?xml version="1.0" encoding="UTF-8"?>
<beans xmlns="http://www.springframework.org/schema/beans"
    xmlns:xsi="http://www.w3.org/2001/XMLSchema-instance"
    xmlns:mvc="http://www.springframework.org/schema/mvc"
    xmlns:context="http://www.springframework.org/schema/context"
    xsi:schemaLocation="
        http://www.springframework.org/schema/beans
        https://www.springframework.org/schema/beans/spring-beans.xsd
        http://www.springframework.org/schema/context
        https://www.springframework.org/schema/context/spring-context.xsd
        http://www.springframework.org/schema/mvc
        https://www.springframework.org/schema/mvc/spring-mvc.xsd">
        <!-- 组件扫描 -->
        <context:component-scan base-package="demoJson.control"></context:component-scan>
        <!-- 注解驱动 -->
        <mvc:annotation-driven/>
        <mvc:resources location="/js/" mapping="/js/**"/>
        <!-- 视图解析器 -->
        <bean class="org.springframework.web.servlet.view.InternalResourceViewResolver">
            <property name="prefix" value="/WEB-INF/jsp/"></property>
            <property name="suffix" value=".jsp"></property>
        </bean>
</beans>
```

以上代码中 mvc:resources 的配置是为了防止静态资源被拦截，本案例中是保证 jquery.js 不被拦截。

(8) 测试：输入信息后单击提交按钮，收到后台回复的 JSON 格式数据，如图 5-19 所示，解析后显示在界面下面。

localhost:8080/WebJson_war_exploded/

注册

用户名: zhangsan

密　码: ••••••

电　话: 187000000

提交

尊敬的zhangsan ,您已注册成功，用户ID为：1710340929382

图 5-19　运行结果

5.5.2　RESTful 实现

RESTful 编程风格是一种软件设计风格，它基于 REST(Representational State Transfer)架构风格，强调使用 HTTP 协议和资源的概念来设计和实现服务。在 RESTful 服务中，一切都是资源，RESTful 风格的 URL 中不得包含任何与操作相关的动词。

当请求中需要携带参数时，RESTFul 允许将参数通过斜杠(/)拼接到 URL 中，而不再像以前一样使用问号(?)拼接键值对的方式来携带参数。例如，访问用户(user)相关的资源时，传统 URL 和 RESTful 的风格对比如下：

```
//传统 URL
http://localhost:8080/demo/queryuser?type=1&id=101
//RESTful 风格 URL
http://localhost:8080/demo/user/1/101
```

RESTful 使用 HTTP 方法对资源执行操作，这些方法具有预定义的含义，如 GET 用于获取资源，POST 用于创建资源，PUT 用于更新或创建资源，DELETE 用于删除资源。

在 Spring MVC 中实现 RESTful 风格的请求，通常使用@RequestMapping 或直接使用@GetMapping、@PostMapping、@PutMapping 和@DeleteMapping 注解处理 GET、POST、PUT 和 DELETE 等 HTTP 请求。

一般通过占位符{xxx}来表示传递的参数，注意占位符的位置应当与请求 URL 中参数的位置保持一致。

使用@PathVariable、@RequestParam 和@RequestBody 等注解来获取请求中的数据。示例代码如下：

```
@GetMapping("/users/{id}")
public User getUserById(@PathVariable("id") Long id) {
    return userService.getUserById(id);
}
@PostMapping("/users")
public User createUser(@RequestBody User user) {
    return userService.createUser(user);
}
@PutMapping("/users/{id}")
public User updateUser(@PathVariable("id") Long id, @RequestBody User user) {
    return userService.updateUser(id, user);
}
@DeleteMapping("/users/{id}")
public String deleteUser(@PathVariable("id") Long id) {
    userService.deleteUser(id);
    return "User deleted";
}
```

5.6 实战演练：驾校学员系统视图层、控制层实现

1. 需求分析

需求分析如下：

序号	需求类别	需 求 描 述
1	视图层需求	视图层分类：首页、学生信息管理、课程管理、考试管理、学习进度、财务分析、档案管理等
		首页：可视化显示科目通过率情况、人员类别通过率情况
		添加学员：允许用户添加学员信息
		搜索学员：允许用户通过名称搜索指定学员
		查询所有学员：显示所有学员的信息
2	控制层需求	可视化数据访问：处理学员通过率数据请求，返回科目通过率、人员类别通过率数据
		学员信息管理：负责处理学员信息的增加、查询请求，返回视图层所需数据

2. 规划设计

填写如下规划设计表：

<table>
<tr><td colspan="4">WBS 表</td></tr>
<tr><td colspan="4">项目基本情况</td></tr>
<tr><td>项目名称</td><td>驾校学员信息系统视图、控制层</td><td>任务编号</td><td></td></tr>
<tr><td>姓　　名</td><td></td><td>班　　级</td><td></td></tr>
<tr><td colspan="4">工作分解</td></tr>
</table>

工作任务	包含活动	备注
1. 视图层开发	1.1 界面顶部导航栏设计	
	1.2 界面底部企业信息栏设计	
	1.3 设计首页界面	
	1.4 设计学生信息管理界面	
	1.5 设计其他界面	
2. 控制层开发	2.1 设计数据可视化访问逻辑	
	2.2 设计学员信息管理控制逻辑	
	2.3 测试	

3. 项目实现

(1) 参照 5.1.2 节的步骤(1)～(8)创建一个名称为 WebStudent 的 Maven 项目，通过 pom.xml 添加如图 5-20 所示的 Spring MVC、JSTL、JSON 等相关依赖包。

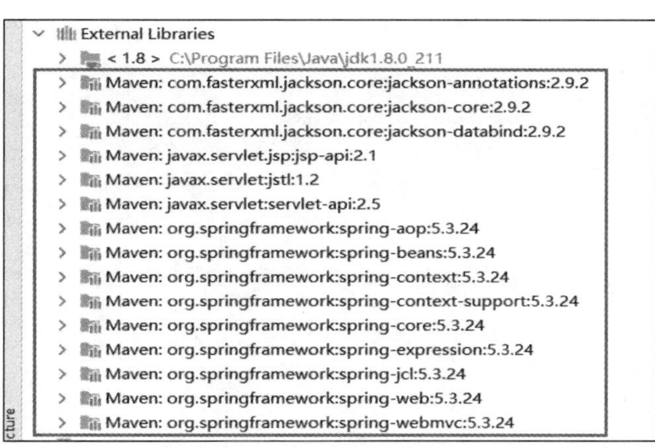

图 5-20　依赖包

(2) 修改 web.xml，对 Spring MVC 前端控制器、编码过滤等进行配置。

web.xml

```
<?xml version="1.0" encoding="UTF-8"?>
<web-app xmlns="http://xmlns.jcp.org/xml/ns/javaee"
        xmlns:xsi="http://www.w3.org/2001/XMLSchema-instance"
        xsi:schemaLocation="http://xmlns.jcp.org/xml/ns/javaee
                            http://xmlns.jcp.org/xml/ns/javaee/web-app_4_0.xsd"
        version="4.0">
    <servlet>
        <!-- 配置前端控制器 -->
        <servlet-name>springmvc</servlet-name>
        <servlet-class>org.springframework.web.servlet.DispatcherServlet</servlet-class>
        <!-- 初始化加载配置文件 -->
        <init-param>
            <param-name>contextConfigLocation</param-name>
            <param-value>classpath:applicationContext.xml</param-value>
        </init-param>
        <!-- 容器启动时立即加载 servlet -->
        <load-on-startup>1</load-on-startup>
    </servlet>
    <!-- 拦截所有的 URL 交前端控制器处理 -->
    <servlet-mapping>
        <servlet-name>springmvc</servlet-name>
        <url-pattern>/</url-pattern>
    </servlet-mapping>
    <!-- 编码过滤器 -->
    <filter>
        <filter-name>myfilter</filter-name>
        <filter-class>org.springframework.web.filter.CharacterEncodingFilter</filter-class>
```

```
            <init-param>
                <param-name>encoding</param-name>
                <param-value>UTF-8</param-value>
            </init-param>
        </filter>
        <filter-mapping>
            <filter-name>myfilter</filter-name>
            <url-pattern>/*</url-pattern>
        </filter-mapping>
        <welcome-file-list>
            <welcome-file>index.jsp</welcome-file>
    </welcome-file-list>
    </web-app>
```

(3) 修改 resources 目录下的 Spring MVC 核心配置文件，完成注解驱动、静态资源处理、视图解析器和控制器扫描等配置。

applicationContext.xml

```
<?xml version="1.0" encoding="UTF-8"?>
<beans xmlns="http://www.springframework.org/schema/beans"
       xmlns:xsi="http://www.w3.org/2001/XMLSchema-instance"
       xmlns:mvc="http://www.springframework.org/schema/mvc"
       xmlns:context="http://www.springframework.org/schema/context"
       xsi:schemaLocation="
             http://www.springframework.org/schema/beans
             https://www.springframework.org/schema/beans/spring-beans.xsd
             http://www.springframework.org/schema/context
             https://www.springframework.org/schema/context/spring-context.xsd
             http://www.springframework.org/schema/mvc
             https://www.springframework.org/schema/mvc/spring-mvc.xsd">
             <!-- 注解驱动 -->
    <mvc:annotation-driven/>
    <!-- 不被拦截的文件  为了防止静态资源被拦截而报错 保证不被拦截-->
    <mvc:resources location="/js/" mapping="/js/**"/>
    <mvc:resources location="/css/" mapping="/css/**"/>
    <!-- 视图解析器 -->
    <bean class="org.springframework.web.servlet.view.InternalResourceViewResolver">
        <property name="prefix" valu="/WEB-INF/jsp/"></property>
        <property name="suffix" value=".jsp"></property>
    </bean>
    <context:component-scan base-package="com.controller"></context:component-scan>
</beans>
```

(4) 创建如图 5-21 所示的控制器包和 POJO 相关包。

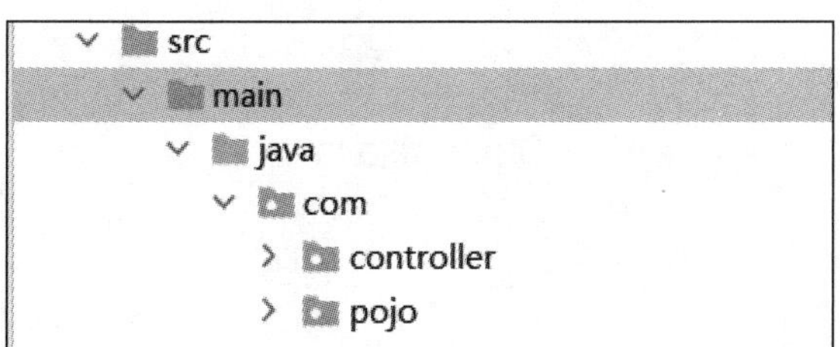

图 5-21　创建包

(5) 创建相关 POJO 类 EchartsData.java 和 Student.java，分别用于存储与 ECharts 图表相关的数据和学生的信息。

EchartsData.java

```
package com.pojo;
import java.util.*;
public class EchartsData {
    public   Map cars ;
    public   Map passRateData ;
    public   List<Map> classPassRate;
    public EchartsData() {
    }
    public   Map getCars() {
        return cars;
    }
    public   void setCars(Map cars) {
        this.cars = cars;
    }
    public   Map getPassRateData() {
        return passRateData;
    }
    public List<Map> getClassPassRate() {
        return classPassRate;
    }
    public void setClassPassRate(List<Map> classPassRate) {
        this.classPassRate = classPassRate;
    }
    public   void setPassRateData(Map passRateData) {
        this.passRateData = passRateData;
    }
    public EchartsData(Map cars, Map passRateData, List<Map> classPassRate) {
        this.cars = cars;
        this.passRateData = passRateData;
```

```
        this.classPassRate=classPassRate;
    }
}
```

Student.java

```
package com.pojo;
public class Student {
    private String studentId;
    private String name;
    private String email;
    private String phone;
    private String address;
    // 构造器
    public Student() {
    }
    public Student(String studentId, String name, String email, String phone, String address) {
        this.studentId = studentId;
        this.name = name;
        this.email = email;
        this.phone = phone;
        this.address = address;
    }
    // Getters 和 Setters
    public String getStudentId() {
        return studentId;
    }
    public void setStudentId(String studentId) {
        this.studentId = studentId;
    }
    public String getName() {
        return name;
    }
    public void setName(String name) {
        this.name = name;
    }
    public String getEmail() {
        return email;
    }
    public void setEmail(String email) {
        this.email = email;
    }
```

```
    public String getPhone() {
        return phone;
    }
    public void setPhone(String phone) {
        this.phone = phone;
    }
    public String getAddress() {
        return address;
    }
    public void setAddress(String address) {
        this.address = address;
    }
    // toString 方法，用于打印学员信息
    @Override
    public String toString() {
        return "Student{" +
                "studentId='" + studentId + '\'' +
                ", name='" + name + '\'' +
                ", email='" + email + '\'' +
                ", phone='" + phone + '\'' +
                ", address='" + address + '\'' +
                '}';
    }
}
```

(6) 在 web 目录下创建如图 5-22 所示的资源文件目录，并导入 echarts.js、jquery-3.3.1.js 库文件，创建样式文件 styles.css 和视图文件 index.jsp。

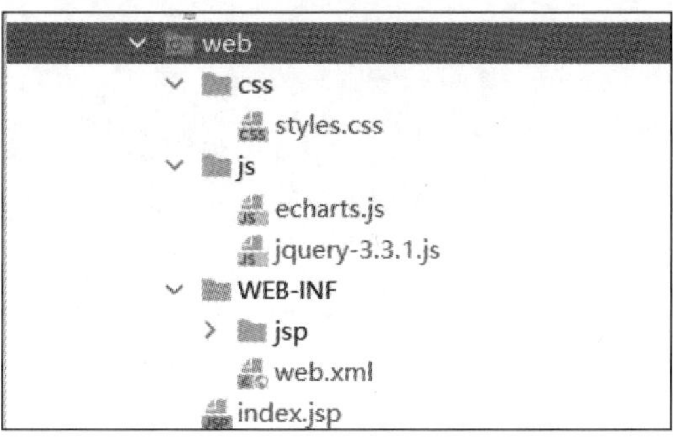

图 5-22　创建资源文件

styles.css

```
/* 设置导航栏的样式 */
#header{
    background-color: #007bff;   /* 深蓝色背景 */
```

```
    color: white;
    display: flex;
    justify-content: space-between;
}
#header h1{
    flex: 1;
    padding: 15px;
}
#header #right{
    flex: 3;
}
#right p{
    width: 100%;
    text-align: center;
    font-size: 25px;
    color: lightgray;
    font-style: italic;
}
#header nav {
    width: 100%;
    display: flex;
    flex-direction: row;
    margin-top: 10px;
    justify-content: space-between;
}
/* 设置导航项的样式 */
nav div{
    flex: 1;
}
/* 设置链接的样式 */
nav div a {
    color: white;
    text-decoration: none;
}
/* 设置当前活动链接的样式 */
nav div a.active {
    background-color: #0056b3; /* 深蓝色背景 */
    color: white;
}
/* 鼠标悬停在链接上的样式 */
```

```
nav div a:hover:not(.active) {
    background-color: #003366; /* 鼠标悬停时的深蓝色背景 */
    color: white; /* 文字保持白色 */
}
table, th, td {
    border: 1px solid black;
}
th, td {
    padding: 5px;
}
#footer{
    background-color: #007bff; /* 深蓝色背景 */
    color: white;
    height: 120px;
}
#footer p{
     line-height: 50px;
     margin-left: 20px;
}
```

index.jsp 文件中实现了向服务器端请求 JSON 数据并使用 ECharts 可视化显示。

index.jsp

```
<%@ page contentType="text/html;charset=UTF-8" language="java" %>
<%@taglib prefix="c" uri="http://java.sun.com/jsp/jstl/core" %>
<html>
<head>
    <title>驾校学员信息管理系统</title>
    <link rel="stylesheet" href="css/styles.css" type="text/css">
    <script src="js/echarts.js"></script>
    <script src="js/jquery-3.3.1.js"></script>
    <style>
        #main1{
            width: 90%;
            height: 400px;
            margin: 20px auto;
            border: 1px black solid;
        }
        #passrate{
            width: 90%;
            height: 420px;
            margin: 20px auto;
```

```
            padding-top: 10px;
            border: 1px black solid;
            text-align: center;
        }
         #main2, #main3{
             width: 45%;
             height: 400px;
             display: inline-block;
        }
    </style>
</head>
<body>
<%@include file="WEB-INF/jsp/head.jsp" %>
<!-- 用于展示图表的容器 -->
<div id="passrate">
    <div id="main2" ></div>
    <div id="main3" ></div>
</div>
<div id="main1"></div>
<%@ include file="WEB-INF/jsp/footer.jsp" %>
<script>
    // 基于准备好的 DOM，初始化 ECharts 实例
    var myChart1 = echarts.init(document.getElementById("main1"))
    var myChart2 = echarts.init(document.getElementById("main2"))
    var myChart3 = echarts.init(document.getElementById("main3"))
    // 绘制图表
    var option1 = {
        title: {
            text: "不同时间段的用车数量",
            x:"center"
        },
        tooltip: {
            trigger: "axis"
        },
        legend: {
            data: ["用车数量"],
            top:"30px",
        },
        grid: {
            left: "3%",
```

```
            right: "4%",
            bottom: "3%",
            containLabel: true
        },
        toolbox: {
            feature: {
                saveAsImage: {}
            }
        },
        xAxis: {
            type: "category",
            boundaryGap: false,
            data: []
        },
        yAxis: {
            type: "value"
        },
        series: [{
            name: "用车数量",
            type: "line",
            data: []
        }]
    };
    // 绘制图表
    var option2 = {
        title: {
            text: "各科通过率",
            x:"center"
        },
        tooltip: {
            trigger: "axis"
        },
        legend: {
            data: ["各科通过率"],
            top:"30px",
        },
        grid: {
            left: "3%",
            right: "4%",
            bottom: "3%",
```

```
            containLabel: true
        },
        toolbox: {
            feature: {
                saveAsImage: {}
            }
        },
        xAxis: {
            type: "category",
            boundaryGap: false,
            //得到关键字数组
            data: []
        },
        yAxis: {
            type: "value"
        },
        series: [{
            name: "各科通过率",
            type: "bar",
            barWidth:"40%",
            //得到关键字数组，然后映射到对应的值数组
            data:[]
        }]
    };
    // 指定图表的配置项和数据
    var option3 = {
        title: {
            text: "不同班级的学员考试通过率",
            x:"center"
        },
        tooltip: {
            trigger: "item"
        },
        legend: {
            top:"30px",
        },
        series: [
            {
                name: "各班型考试通过率",
                type: "pie",
```

```
                radius: ["80%", "30%"],
                center: ["50%", "60%"],
                data: [],
                emphasis: {
                    itemStyle: {
                        shadowBlur: 10,
                        shadowOffsetX: 0,
                        shadowColor: "rgba(0, 0, 0, 0.5)"
                    }
                }
            } ]
    };
    $(document).ready(function () {
        $.ajax({
            type: "GET",
            url: "${pageContext.request.contextPath}/report",
            dataType:"json",
            success: function (res) {
                console.log(res)
                var cars = res.cars
                var passRateData = res.passRateData
                var classPassRate = res.classPassRate
                option1.xAxis.data = cars.names
                option1.series[0].data = cars.values
                myChart1.setOption(option1);
                option2.xAxis.data = passRateData.names
                option2.series[0].data = passRateData.values
                myChart2.setOption(option2);
                console.log(classPassRate)
                option3.series[0].data = classPassRate
                myChart3.setOption(option3);
            }
        })
    })
</script>
</body>
</html>
```

(7) 在 jsp 目录下创建其他页面文件 student.jsp、head.jsp 和 footer.jsp，其中 student.jsp 实现向服务器端提交添加的学生信息，并从解析服务器端回传学校信息列表。

student.jsp

```
<%@ page contentType="text/html; charset=UTF-8" language="java" %>
<%@taglib prefix="c" uri="http://java.sun.com/jsp/jstl/core" %>
<html>
<head>
    <title>学员信息管理</title>
    <link rel="stylesheet" href="${pageContext.request.contextPath}/css/styles.css"
    " type = text/css">
    <script src="${pageContext.request.contextPath}/js/jquery-3.3.1.js"></script>
    <style>
        form{
            margin:20px;
        }
        #liststu{
            margin: 20px;
        }
    </style>
</head>
<body>
<%@include file="head.jsp"%>
<%-- 学员信息表单 --%>
<form action="${pageContext.request.contextPath}/addstudent" method="post">
    <h2>-------添加学员-------</h2>
    <div>
        <input type="hidden" id="studentId" name="studentId" readonly="readonly" />
    </div>
    <div>
        <label for="name">姓名：</label>
        <input type="text" id="name" name="name" required="required" />
    </div>
    <div>
        <label for="email">邮箱：</label>
        <input type="email" id="email" name="email" required="required" />
    </div>
    <div>
        <label for="phone">电话：</label>
        <input type="tel" id="phone" name="phone" required="required" />
    </div>
    <div>
        <label for="address">地址：</label>
```

```
            <input type="text" id="address" name="address" />
        </div>
        <div>
            <input type="submit" value="保存"/>
        </div>
    </form>
    <div id="liststu">
        <h2>-------学员列表-------</h2>
        <div>
            <input type="text" id="search" placeholder="搜索学员" />
            <input type="button" id="mybutton" value="搜索"/>
        </div>
        <table>
            <thead>
            <tr>
                <th>学员 ID</th>
                <th>姓名</th>
                <th>邮箱</th>
                <th>电话</th>
                <th>地址</th>
            </tr>
            </thead>
            <tbody>
            <%-- 在这里添加学员信息的行 --%>
            <c:forEach var="student" items="${studentlist}" varStatus="count">
                <tr>
                    <td>${student.studentId}</td>
                    <td>${student.name}</td>
                    <td>${student.email}</td>
                    <td>${student.phone}</td>
                    <td>${student.address}</td>
                </tr>
            </c:forEach>
            </tbody>
        </table>
    </div>
    <%@include file="footer.jsp"%>
    <script>
        $("#mybutton").click(function (){
            var search=$("#search").val().trim()
```

```
            if(search){
                var result= $('tr:contains("'+search+'")')
                if(result.html()){
                    console.log(result.html())
                    $('tr:not(:contains("'+search+'"))').hide()
                }
                else
                    location.reload()
            }
            else
                location.reload()
        })
    </script>
    </body>
    </html>
```

head.jsp

```
<%@ page contentType="text/html;charset=UTF-8" language="java" %>
<div id="header">
    <h1>驾校学员管理系统</h1>
    <div id="right">
        <p>握紧方向盘      成就未来    安全驾驶
              从我们开始 </p>
        <nav>
            <div><a href="${pageContext.request.contextPath}/">首页</a></div>
            <div><a href="${pageContext.request.contextPath}/student">学员信息</a></div>
            <div><a href="#">课程管理</a></div>
            <div><a href="#">考试管理</a></div>
            <div><a href="#">学习进度</a></div>
            <div><a href="#">财务分析</a></div>
            <div><a href="#">档案管理</a></div>
            <div><a href="#">车辆安全</a></div>
        </nav>
    </div>
</div>
```

footer.jsp

```
<%@ page contentType="text/html; charset=UTF-8" language="java" %>
<div id="footer"   class="tb">
    <p>&copy; 2023 驾校学员管理系统. 版权所有.</p>
    <p>联系我们：  studentinfo@driving-school.com</p>
</div>
```

(8) 在 controller 包下创建控制器类 Mycontroller.java，实现对页面请求的处理。

Mycontroller.java

```
package com.controller;
import com.pojo.EchartsData;
import com.pojo.Student;
import org.springframework.stereotype.Controller;
import org.springframework.ui.Model;
import org.springframework.web.bind.annotation.GetMapping;
import org.springframework.web.bind.annotation.PostMapping;
import org.springframework.web.bind.annotation.RequestMapping;
import org.springframework.web.bind.annotation.ResponseBody;
import java.util.*;
@Controller
public class Mycontroller {
  public   static List<Student> studentList=new ArrayList<>();
  static   {
        String[] names = {"张晓东", "李大壮", "王晓敏"};
        String[] emails  =  {"zhangxxxdong@qq.com",  "lidaxxx@163.com",  "wangxxxmin@
139.com"};
        String[] phones = {"029-xxxxxxx", "0916-xxxxx", "139xxxxxxxx"};
        String[] address  =  {"西安市 xxx 路陕西 xxx 学院", "宝鸡市 xxx 路", "延安市 xxx 路 xxx
号"};
        for (int i=0; i<names.length; i++){
            Student student=new Student("c"+(i+1), names[i], emails[i], phones[i], address[i]);
            studentList.add(student);
        }
  }
    //首页
    @GetMapping("/report")
    public   @ResponseBody EchartsData getData(){
        Map   cars = new HashMap<String, Integer>();
        Map   passRateData = new HashMap<String, Integer>();
        List<Map>   classPassRate = new ArrayList<Map>();
        Random r=new Random();
        List names=new ArrayList();
        List values=new ArrayList();
        for(int i=8; i<=18; i++){
            names.add(i+":00");
```

```
                values.add(r.nextInt(80)+20);
            }
            cars.put("names", names);
            cars.put("values", values);
            names=new ArrayList();
            values=new ArrayList();
            for(int i=0; i<5; i++){
                names.add("Course "+(char)('A'+i));
                values.add(r.nextInt(100));
                Map map=new HashMap();
                map.put("value", Math.ceil(100*Math.random())/100);
                map.put("name", "class"+(char)('A'+i));
                classPassRate.add(map);
            }
            passRateData.put("names", names);
            passRateData.put("values", values);
            EchartsData echartsData=new EchartsData(cars, passRateData, classPassRate);
            return   echartsData;
        }
        //学员管理
        @RequestMapping("/student")
        public String getStudent(Model model){
            model.addAttribute("studentlist", studentList);
            return   "student";
        }
        @PostMapping("/addstudent")
        public String addStudent(Model model, Student student){
            student.setStudentId("c"+(studentList.size()+1));
            studentList.add(student);
            model.addAttribute("studentlist", studentList);
            return   "redirect:/student";
        }
    }
```

(9) 部署项目到 tomcat 后运行，首页面从服务器端获取数据并可视化显示，结果如图 5-23 所示，其中显示了科目通过率情况、人员类别通过率等情况。

单击学员信息，进入如图 5-24 所示的学员信息管理页面；搜索和添加学员信息，向服务器端发送学员信息，服务器端回传最新的如图 5-25 所示的学员信息列表并显示在下方。至此，所有视图层和控制层所涉及的功能需求都已经完成。

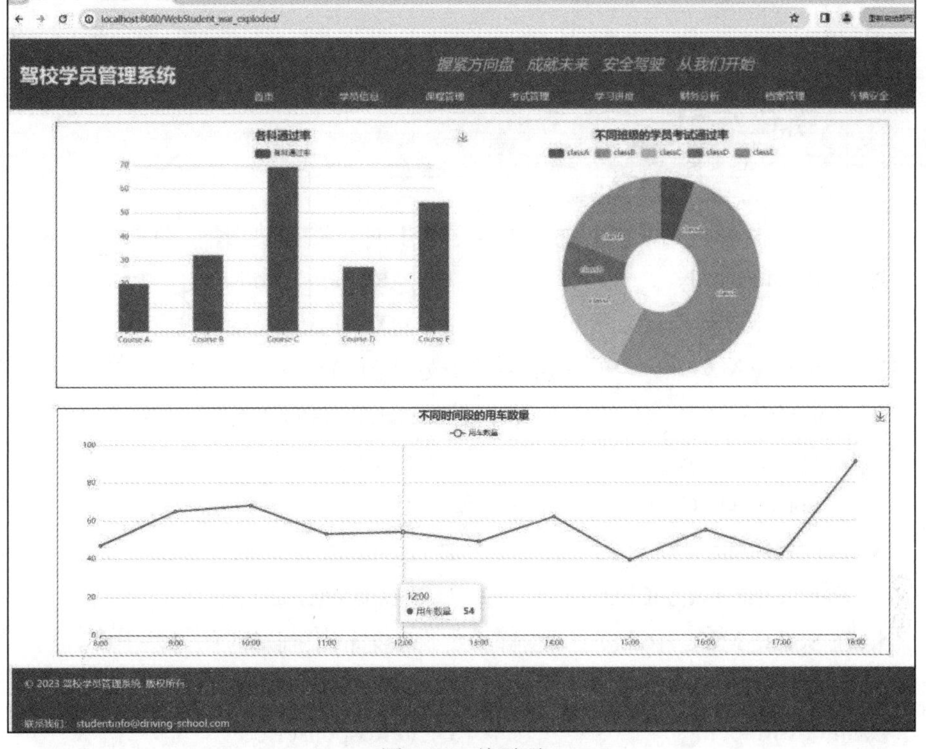

图 5-23　首页面

驾校学员管理系统

首页　学员信息

--------添加学员-------

姓名：李明
邮箱：1234@qq.com
电话：13200000000
地址：北京市
保存

图 5-24　添加学员

-------学员列表-------

搜索学员　搜索

学员ID	姓名	邮箱	电话	地址
c1	张晓东	zhangxxxdong@qq.com	029-xxxxxxx	西安市xxx路陕西xxx学院
c2	李大壮	lidaxxx@163.com	0916-xxxxx	宝鸡市xxx路
c3	王晓敏	wangxxxmin@139.com	139xxxxxxxx	延安市xxx路xxx号
c4	李明	1234@qq.com	13200000000	北京市

© 2023 驾校学员管理系统. 版权所有.

联系我们： studentinfo@driving-school.com

图 5-25　学员列表显示

习题答案

强化练习

1. 当需要将 Java 对象转换为 JSON 字符串时，通常使用(　　)注解。

A. @RequestBody　　B. @ResponseBody
C. @JsonProperty　　D. @XmlRootElement

2. 在 Spring MVC 中，(　　)组件用于处理 HTTP 请求。

A. Controller　　B. HandlerAdapter
C. ModelAndView　　D. HttpServlet

3. 在 Spring MVC 中，(　　)类负责管理所有的 Controller。

A. ServletContext　　B. HandlerMapping
C. HandlerExecutionChain　　D. DispatcherServlet

4. 在 Spring MVC 中，(　　)注解不负责映射请求到对应的方法上。

A. @RequestMapping　　B. @GetMapping
C. @PostMapping　　D. @ResponseBody

5. 在 Spring MVC 中，(　　)注解用于将方法参数标记为请求参数。

A. @RequestParam　　B. @PathVariable
C. @RequestBody　　D. @ResponseBody

6. 在 Restful Web 服务中，(　　)方法用于更新资源。

A. GET　　B. POST
C. PUT　　D. DELETE

7. 在 Spring MVC 中，(　　)类负责将请求映射到 Controller 方法。

A. HandlerMapping　　B. HandlerAdapter
C. HandlerExecutionChain　　D. DispatcherServlet

进一步学习建议

学习完 Spring MVC 的基础之后，可以继续深入学习以下内容：

(1) 学习使用 JSTL、EL 表达式和国际化。

(2) 学习使用 Spring MVC 进行文件上传和下载操作。

(3) 深入了解 Spring MVC 中的数据转换服务(Conversion Service)和格式化服务(Formatter)，学习将服务层的方法调用结果转换为不同的格式，如 JSON、XML 等。

考核评价

考核评价表				
姓名		班　级		
学号		考评时间		
评价主题及总分		评价内容及分数		评分
1	知识考核(30)	阐述 Spring MVC 的核心组件及其作用(10 分)		
		举例说明 Spring MVC 接收请求数据的常见方式(10 分)		
		说明 Spring MVC 应用程序处理 JSON 格式的请求和响应的方法(10 分)		
2	技能考核(40)	具备业务需求分析、功能设计、编码及测试的综合能力(10 分)		
		开发任务能够按时完成(20 分)		
		熟练编写控制器方法，实现绑定不同数据类型的形参和 JSON 数据转换(10 分)		
3	思政考核(30)	收集在技术进步中积极履行社会责任、推动社会发展的成功案例(10 分)		
		介绍一个开源项目并阐述开放共享的重要性(10 分)		
		进行反思和总结，分析自己在编程能力和解决问题方面的优点和不足(10 分)		
评语：				汇总：

第6章　MyBatis框架

学习目标

目标类型	目 标 描 述
知识目标	• 理解MyBatis的基本概念和原理 • 掌握MyBatis的核心配置文件和映射文件的结构 • 掌握MyBatis的注解方式和XML方式的配置和使用，了解两者之间的区别和优劣势 • 掌握MyBatis动态SQL语句的编写和使用
技能目标	• 熟练使用MyBatis进行数据库操作，包括增删改查等基本操作 • 编写高质量MyBatis映射文件，实现数据库表与Java对象映射 • 使用MyBatis的注解方式或XML方式进行配置，灵活切换不同的配置方式
思政目标	• 养成良好的学习习惯和自学能力，主动学习和掌握新的知识与技能 • 提高解决问题的能力，能够独立分析和解决遇到的问题 • 培养团队合作意识，能够与他人合作完成项目，共同解决技术难题 • 提高沟通能力，能够清晰地表达自己的想法和观点，与他人进行有效的沟通和交流

知识技能储备

6.1　MyBatis快速上手

6.1.1　MyBatis简介

MyBatis是一个基于Java的持久层框架，旨在简化数据库操作并提高代码的可读性和可维护性，其核心功能是将Java对象与数据库表中的记录进行映射，从而简化了开发者的工作；MyBatis支持多种映射配置，如XML配置文件、注解等，可以满足不同开发需求。

6.1.2　入门指南

MyBatis 入门

为了帮助读者快速掌握 MyBatis 的使用，下面通过一个入门程序来演示编程过程，其具体步骤如下：

(1) 准备数据源：创建如图 6-1 所示脚本文件，实现创建 student 表并插入两条记录。

```
data.sql
DROP DATABASE IF EXISTS test;
CREATE DATABASE test;
USE  test;
CREATE TABLE student(
  id  INT  NOT NULL AUTO_INCREMENT,
  age INT,
  name VARCHAR(20),
  sex VARCHAR(20),
  PRIMARY KEY(id)
  );
 INSERT INTO student VALUES(NULL,32,'李明','男');
 INSERT INTO student VALUES(NULL,33,'王伟','女');
```

图 6-1　脚本文件截图

(2) 输入指令，执行上图中的 SQL 脚本内容，具体指令如下：

```
$ mysql -uroot -p
Enter password: ****
Welcome to the MySQL monitor.  Commands end with ; or \g.
mysql> source data.sql
Query OK, 1 row affected (0.04 sec)
Query OK, 1 row affected (0.00 sec)
Database changed
Query OK, 0 rows affected (0.01 sec)
Query OK, 1 row affected (0.00 sec)
Query OK, 1 row affected (0.00 sec)
mysql>
```

以上命令执行结果说明数据库脚本执行正常。

(3) 按照如图 6-2 所示创建 Maven 项目，再参照图 6-3 所示输入项目名称。

图 6-2　选择 Maven 项目

New Project

Name: FirstMybatis

Location: D:\EChartsPrj\mybooksource\FirstMybatis

Artifact Coordinates

图 6-3　输入项目名称

(4) 在 pom.xml 文件中添加 mybatis 和 mysql 依赖。

pom.xml

```
<?xml version="1.0" encoding="UTF-8"?>
<project xmlns="http://maven.apache.org/POM/4.0.0"
         xmlns:xsi="http://www.w3.org/2001/XMLSchema-instance"
         xsi:schemaLocation="http://maven.apache.org/POM/4.0.0
                             http://maven.apache.org/xsd/maven-4.0.0.xsd">
    <modelVersion>4.0.0</modelVersion>
    <groupId>org.example</groupId>
    <artifactId>FirstMybatis</artifactId>
    <version>1.0-SNAPSHOT</version>
    <properties>
        <maven.compiler.source>8</maven.compiler.source>
        <maven.compiler.target>8</maven.compiler.target>
    </properties>
    <dependencies>
        <dependency>
            <groupId>org.mybatis</groupId>
            <artifactId>mybatis</artifactId>
            <version>3.5.10</version>
        </dependency>
        <dependency>
            <groupId>mysql</groupId>
            <artifactId>mysql-connector-java</artifactId>
            <version>5.1.49</version>
        </dependency>
    </dependencies>
    <!--在 build 中配置 resources，来防止资源导出失败的问题-->
    <build>
        <resources>
            <resource>
                <directory>src/main/resources</directory>
                <includes>
                    <include>**/*.properties</include>
                    <include>**/*.xml</include>
                </includes>
                <filtering>true</filtering>
            </resource>
            <resource>
                <directory>src/main/java</directory>
                <includes>
                    <include>**/*.properties</include>
```

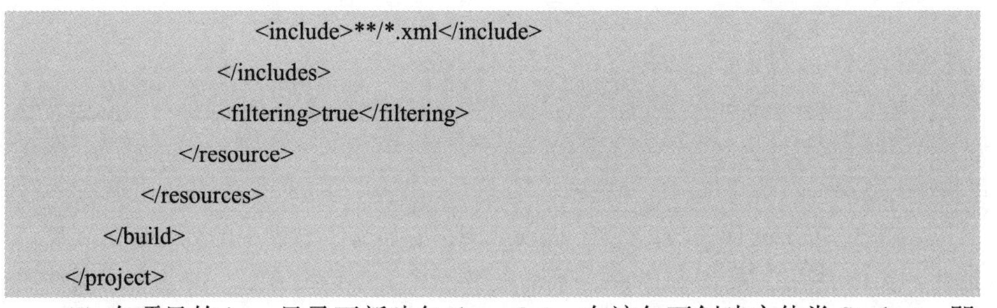

```
                    <include>**/*.xml</include>
                </includes>
                <filtering>true</filtering>
            </resource>
        </resources>
    </build>
</project>
```

(5) 在项目的 java 目录下新建包 demo6_1，在该包下创建实体类 Student，即 Student.java。

Student.java

```
package demo6_1;
public class Student {
    int id;
    int age;
    String name;
    String gender;
    public int getId() {
        return id;
    }
    public void setId(int id) {
        this.id = id;
    }
    public int getAge() {
        return age;
    }
    public void setAge(int age) {
        this.age = age;
    }
    public String getName() {
        return name;
    }
    public void setName(String name) {
        this.name = name;
    }
    public String getGender() {
        return gender;
    }
    public void setGender(String gender) {
        this.gender = gender;
    }
```

```
    @Override
    public String toString() {
        return "Student [id=" + id + ", age=" + age + ", name=" + name + ", gender=" + gender + "]";
    }
}
```

实体类 Student 代表数据库中 student 表，它包含了与表中的列对应的属性，通常，实体类需遵循 Java Bean 规范，提供一个无参构造方法，并为每个属性提供 getter 和 setter 方法。

(6) 创建映射文件 map.xml 用于定义 SQL 语句，从官网拷贝模板后按照以下内容修改：

```
<?xml version="1.0" encoding="UTF-8" ?>
<!DOCTYPE mapper
  PUBLIC "-//mybatis.org//DTD Mapper 3.0//EN"
  "http://mybatis.org/dtd/mybatis-3-mapper.dtd">
<mapper namespace="demo6_1.map">
  <select id="selectStudent" resultType="demo6_1.Student" parameterType="Integer">
    select * from student where id=#{id}
  </select>
</mapper>
```

映射文件用于定义 SQL 语句，每个文件对应数据库的一张表；本例在命名空间“demo6_1.map”中定义一个名为“selectStudent”的映射语句，这样 SqlSession 就可以用全限定名“demo6_1.map.selectStudent”来调用映射语句；其中 resultType 表示结果为 Student 类型，从而实现 Java 对象与数据库操作结果类型的匹配。

(7) 在 resources 目录下创建 MyBatis 核心配置文件 mybatis-config.xml，可从官网拷贝模板再进行内容修改，最终该配置文件内容如下：

```
<?xml version="1.0" encoding="UTF-8" ?>
<!DOCTYPE configuration
        PUBLIC "-//mybatis.org//DTD Config 3.0//EN"
        "http://mybatis.org/dtd/mybatis-3-config.dtd">
<configuration>
    <settings>
        <setting name="logImpl" value="STDOUT_LOGGING" />
    </settings>
    <environments default="development">
        <environment id="development">
            <transactionManager type="JDBC"/>
            <dataSource type="POOLED">
                <property name="driver" value="com.mysql.jdbc.Driver"/>
                <property name="url" value="jdbc:mysql://localhost:3306/test"/>
                <property name="username" value="root"/>
```

```
                <property name="password" value="root"/>
            </dataSource>
        </environment>
    </environments>
    <mappers>
        <mapper resource="demo6_1/map.xml"/>
    </mappers>
</configuration>
```

以上文件包含对 MyBatis 系统的核心设置：获取数据库连接实例的数据源(DataSource)以及事务管理器(TransactionManager)。<dataSource>标签表示数据源；type="POOLED" 表示使用连接池技术；<property>标签用于设置数据库连接的相关属性，包括驱动类名(driver)、数据库地址(url)、用户名(username)和密码(password)。<mappers>元素则包含一组映射器<mapper>，这些映射器的 resource 属性对应上一步定义的映射文件。

注意事项：Mybatis 3.5 集成控制台 log 的功能，不再需要引入 log4j；只需要在核心配置文件中添加 logImpl 设置，即可开启 SQL 查询日志的输出。

(8) 创建测试类 Test.java，需要在测试类中使用 SqlSessionFactory 和 SqlSession 来操作数据库，具体方法如下：

```
package demo6_1;
import java.io.IOException;
import java.io.InputStream;
import org.apache.ibatis.io.Resources;
import org.apache.ibatis.session.SqlSession;
import org.apache.ibatis.session.SqlSessionFactory;
import org.apache.ibatis.session.SqlSessionFactoryBuilder;
public class Test {
    public static void main(String[] a) {
        String resourseString="mybatis-config.xml";
        try {
            InputStream inputStream=Resources.getResourceAsStream(resourseString);
            SqlSessionFactory sessionFactory=new SqlSessionFactoryBuilder().
            build(inputStream);
            SqlSession session=sessionFactory.openSession();
            Student student=(Student)session.selectOne("selectStudent", 1);
            System.out.println(student.toString());
        } catch (IOException e) {
            e.printStackTrace();
        }
    }
}
```

SqlSessionFactory 是 MyBatis 框架中十分重要的对象，其主要作用是创建 SqlSession，而 SqlSession 是应用程序与持久层之间执行交互操作的一个单线程对象，提供在数据库执行 SQL 命令所需的所有方法，其主要作用是执行持久化操作。

以上代码通过 Resources.getResourceAsStream(resourseString)方法获取到配置文件的输入流。使用 SqlSessionFactoryBuilder 类的 build 方法构建一个 SqlSessionFactory 对象，通过该对象的 openSession 方法开启一个 SqlSession 类型的会话。使用 SqlSession 的 selectOne 方法，并传入 SQL 语句 ID("selectStudent")和查询参数(1)，查询 id 为 1 的学生信息，并将结果转换为 Student 类型。

(9) 运行 Test.java 文件的结果如下：

```
==>  Preparing: select * from student where id=?
==> Parameters: 1(Integer)
<==      Columns: id, age, name, gender
<==          Row: 1, 32, 李明, 男
<==        Total: 1    Student [id=1, age=32, name=李明, gender=男]
```

通过该结果发现使用 Mybatis 框架成功地获取了数据。

6.2 配 置 文 件

MyBatis 的配置文件包含了影响 MyBatis 行为的设置和属性信息，配置文档的层次结构如图 6-4 所示。

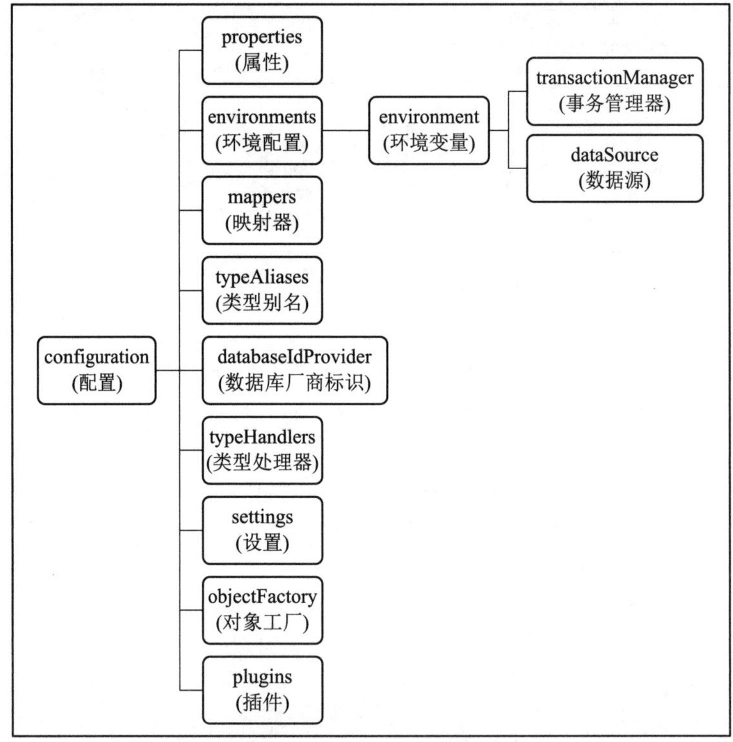

图 6-4 配置文件结构

从图 6-4 可以清晰地看到 Mybatis 配置文件结构，本节将介绍该文件的各部分作用和具体配置。

6.2.1　属性(properties)

元素<properties>代表了属性配置，既可以在外部的 Java 属性文件进行配置，也可以在标签内部动态替换。示例代码如下：

```
<properties resource="org/mybatis/example/config.properties">
    <property name="username" value="dev_user"/>
    <property name="password" value="F2Fa3!33TYyg"/>
</properties>
```

上述代码中的 username 和 password 属性将由<property>元素设置。而 driver 和 url 属性将在 config.properties 文件中配置。如下 MyBatis 配置实现了由 config.properties 动态替换数据库连接属性：

```
<dataSource type="POOLED">
    <property name="driver" value="${driver}"/>
    <property name="url" value="${url}"/>
    <property name="username" value="${username}"/>
    <property name="password" value="${password}"/>
</dataSource>
```

6.2.2　环境配置(environments)

1. MyBatis 的环境概述

元素<environments>定义了配置环境的方式，MyBatis 可以通过配置来适应多种环境，这种机制有助于将 SQL 映射应用于多种数据库之中。例如在开发、测试和生产环境实现有不同的配置；或者在具有相同 Schema 的多个数据库中使用相同的 SQL 映射。

environments 元素定义配置环境的方式如下：

```
<environments default="development">
    <environment id="development">
        <transactionManager type="JDBC">
            <property name="..." value="..."/>
        </transactionManager>
        <dataSource type="POOLED">
            <property name="driver" value="${driver}"/>
            <property name="url" value="${url}"/>
            <property name="username" value="${username}"/>
            <property name="password" value="${password}"/>
        </dataSource>
    </environment>
</environments>
```

尽管可以配置多个环境，但每个 SqlSessionFactory 实例只能选择一种环境。假设想连接两个数据库，就需要创建两个 SqlSessionFactory 实例，每个数据库对应一个 SqlSessionFactory 实例。为指定创建哪种环境，只需将它作为可选的参数传递给 SqlSessionFactoryBuilder 即可，示例代码如下：

```
// 读取 MyBatis 配置文件
InputStream inputStream = Resources.getResourceAsStream("mybatis-config.xml");
// 创建 SqlSessionFactory
SqlSessionFactory sqlSessionFactory = new SqlSessionFactoryBuilder().build(inputStream,
"development");
```

2. 事务管理器

在 MyBatis 中有两种类型的事务管理器，具体如下：

(1) JDBC：表示使用 JDBC 的提交和回滚功能，它依赖从数据源获得的连接来管理事务作用域。默认情况下，为了与某些驱动程序兼容，它在关闭连接时会启用自动提交。对于某些驱动程序来说，启用自动提交不仅是不必要的，而且是一个代价高昂的操作。因此从 3.5.10 版本开始，可以通过将 "skipSetAutoCommitOnClose" 属性设置为 "true" 来跳过这个步骤，具体代码如下：

```
<transactionManager type="JDBC">
<property name="skipSetAutoCommitOnClose" value="true"/>
</transactionManager>
```

(2) MANAGED：表示从不提交或回滚一个连接，而是让容器来管理事务的整个生命周期。默认情况下它会关闭连接，然而一些容器并不希望连接被关闭，因此需要将 closeConnection 属性设置为 "false" 来阻止默认的关闭行为，具体代码如下：

```
<transactionManager type="MANAGED">
<property name="closeConnection" value="false"/>
</transactionManager>
```

3. dataSource 数据源

元素<dataSource>用于配置数据源，使用标准的 JDBC 数据源接口来配置 JDBC 连接对象的资源，内建的数据源类型有以下 3 种：

(1) UNPOOLED 数据源。每次请求时打开和关闭连接，其性能表现依赖于使用的数据库。虽然速度慢，但对简单应用程序来说是一个很好的选择。UNPOOLED 类型数据源的常见属性如下：

① driver：JDBC 驱动的 Java 类全限定名，必选项。

② url：数据库的 JDBC URL 地址，必选项。

③ username：登录数据库的用户名，必选项。

④ password：登录数据库的密码，必选项。

⑤ defaultTransactionIsolationLevel：默认的事务隔离级别，必选项。

⑥ defaultNetworkTimeout：数据库操作默认超时时间，可选项。

(2) POOLED 数据源。利用“池”的概念将 JDBC 连接对象组织起来，避免创建新的连接实例时所必需的初始化和认证时间，使并发 Web 应用快速响应请求。POOLED 除了具备 UNPOOLED 的属性外，还有以下其他属性：

① poolMaximumActiveConnections：表示最大活动(正在使用)连接数量，默认值为 10。

② poolMaximumIdleConnections：表示最大空闲连接数。

③ poolMaximumCheckoutTime：表示一个连接在池中被借出的最长时间，防止连接被长时间占用而不归还，默认值为 20 000 ms。

④ poolTimeToWait：表示尝试获取一个可用连接时，愿意等待的时间，默认值为 20 000 ms。

⑤ poolPingConnectionsNotUsedFor：表示在指定连接池内没有使用连接时，应该进行 ping 操作检查的时间长度。

(3) JNDI 数据源。可在 EJB 或应用服务器这类容器中被使用，具体属性如下：

① initial_context：用于指定初始上下文。

② data_source：表示引用数据源实例位置的上下文路径。

6.2.3 映射器(mappers)

映射器的主要作用是提供 MyBatis 获取 SQL 映射语句的路径，包括类路径的资源引用、完全限定资源定位符、完全限定类名和包名 4 种形式，示例代码如下：

```
<!-- 使用相对于类路径的资源引用 -->
<mappers>
    <mapper resource="org/mybatis/builder/AuthorMapper.xml"/>
    <mapper resource="org/mybatis/builder/BlogMapper.xml"/>
    <mapper resource="org/mybatis/builder/PostMapper.xml"/>
</mappers>
<!-- 使用完全限定资源定位符(URL) -->
<mappers>
    <mapper url="file:///var/mappers/AuthorMapper.xml"/>
    <mapper url="file:///var/mappers/BlogMapper.xml"/>
    <mapper url="file:///var/mappers/PostMapper.xml"/>
</mappers>
<!-- 使用映射器接口实现类的完全限定类名 -->
<mappers>
    <mapper class="org.mybatis.builder.AuthorMapper"/>
    <mapper class="org.mybatis.builder.BlogMapper"/>
    <mapper class="org.mybatis.builder.PostMapper"/>
</mappers>
<!-- 将包内的映射器接口实现全部注册为映射器 -->
<mappers>
    <package name="org.mybatis.builder"/>
</mappers>
```

6.2.4 类型别名(typeAliases)

类型别名可为 Java 类型设置一个缩写名字，它仅用于 XML 配置，意在降低冗余的全限定类名书写。例如，当在任何有 domain.blog.Blog 的地方使用直接使用 Blog 时，可以进行如下配置：

```
<typeAliases>
    <typeAlias alias="Author" type="domain.blog.Author"/>
    <typeAlias alias="Blog" type="domain.blog.Blog"/>
    <typeAlias alias="Comment" type="domain.blog.Comment"/>
    <typeAlias alias="Post" type="domain.blog.Post"/>
    <typeAlias alias="Section" type="domain.blog.Section"/>
    <typeAlias alias="Tag" type="domain.blog.Tag"/>
</typeAliases>
```

也可以指定一个包名，MyBatis 会在包名下面搜索需要的 Java Bean。例如，配置 domain.blog 包中的每一个 Java Bean，在没有注解的情况下，会使用 Bean 的首字母小写的非限定类名来作为它的别名，即 domain.blog.Author 的别名为 author，具体代码如下：

```
<typeAliases>
    <package name="domain.blog"/>
</typeAliases>
```

6.2.5 其他部分

1. 设置(settings)

设置映射器的缓存、延迟加载、列标签代替列名等，详见 Mybatis 官方文档。

2. 类型处理器(typeHandlers)

MyBatis 在设置预处理语句(PreparedStatement)中的参数或从结果集中取出一个值时，都会用类型处理器将获取到的值以合适的方式转换成 Java 类型。可通过重写已有的类型处理器或创建自己的类型处理器来处理不支持的或非标准的类型，即实现 org.apache.ibatis.type.TypeHandler 接口或继承 org.apache.ibatis.type.BaseTypeHandler 类，并将它映射到一个 JDBC 类型，然后按照如下方式在 mybatis-config.xml 注册类处理器：

```
<typeHandlers>
    <typeHandler handler="org.mybatis.example.ExampleTypeHandler"/>
</typeHandlers>
```

3. 对象工厂(objectFactory)

MyBatis 每次创建结果对象的新实例时，它都会使用默认的对象工厂实例来完成实例化工作。如果要覆盖对象工厂的默认行为，可以通过创建自己的对象工厂类来实现，然后在<objectFactory>中注册自定义对象工厂。

4. 插件(plugins)

MyBatis 允许使用插件在映射语句执行过程中进行拦截调用，详见 MyBatis 官方文档。

5. 数据库厂商标识(databaseIdProvider)

MyBatis 映射语句中的 databaseId 属性表示数据库厂商，为支持多厂商特性，只需在 mybatis-config.xml 文件中加入<databaseIdProvider>即可，具体代码如下：

```
<databaseIdProvider type="DB_VENDOR">
    <property name="SQL Server" value="sqlserver"/>
    <property name="DB2" value="db2"/>
    <property name="Oracle" value="oracle" />
</databaseIdProvider>
```

type 属性指定 ID provider 的类型，通常是 DB_VENDOR，这意味着 MyBatis 会从上下文变量中获取数据库供应商的名称。<property>元素定义数据库 ID 和对应数据库供应商名称之间的映射关系。

6.3　MyBatis 常见 API

SqlSessionFactory 的主要作用是创建 SqlSession，而 SqlSession 是应用程序与持久层之间执行交互操作的一个单线程对象，其主要作用是执行持久化操作。Sql SessionFactory 是由 SqlSessionFactoryBuilder 创建的，可以从 XML、注解或 Java 配置代码来创建 SqlSessionFactory。

6.3.1　创建实例

SqlSessionFactoryBuilder 有 5 种 build 方法，实现从不同的资源中创建一个 Sql SessionFactory 实例，具体方法如下：

```
SqlSessionFactory build(InputStream inputStream)
SqlSessionFactory build(InputStream inputStream, String environment)
SqlSessionFactory build(InputStream inputStream, Properties properties)
SqlSessionFactory build(InputStream inputStream, String env, Properties props)
SqlSessionFactory build(Configuration config)
```

上述前 4 种方法最常见，实现了从包含 MyBatis 配置信息的 XML 文件中创建 SqlSessionFactory。其中：inputStream 参数指向 XML 文件(即 6.1.2 节的 mybatis-config.xml 文件)；environment 参数决定加载哪种环境，如果调用不带 environment 参数的 build 方法，那么就会使用默认的环境配置(即 default 属性)，详见 6.2.2 节环境配置；properties 参数表示配置属性，可用于覆盖配置文件(通常是 mybatis-config.xml)中的属性。

最后一种方法使用已经创建并配置好的 Configuration 对象来构建 SqlSessionFactory。这种方式适用于完全以编程方式创建和配置 MyBatis 的场景，

而不使用 XML 配置文件。

SqlSessionFactory 有多个方法创建 SqlSession 实例，有已重载的多种 openSession 方法供使用，具体如下：

```
SqlSession openSession()
SqlSession openSession(boolean autoCommit)
SqlSession openSession(Connection connection)
SqlSession openSession(TransactionIsolationLevel level)
SqlSession openSession(ExecutorType execType, TransactionIsolationLevel level)
SqlSession openSession(ExecutorType execType)
SqlSession openSession(ExecutorType execType, boolean autoCommit)
SqlSession openSession(ExecutorType execType, Connection connection)
Configuration getConfiguration();
```

默认的 openSession 方法没有参数，它会创建具备如下特性的 SqlSession：

(1) 事务作用域将会开启，即不自动提交。

(2) 将从当前环境配置的 DataSource 实例中获取 Connection 对象。

(3) 事务隔离级别将会使用驱动或数据源的默认设置。

(4) 预处理语句不会被复用，也不会批量处理更新。

其他几种 openSession 方法中涉及的参数为：autoCommit 参数用于指定是否自动提交事务，connection 参数表示使用一个已经存在的数据库连接，level 参数用于指定事务的隔离级别，execType 参数用于指定执行器类型。

getConfiguration 方法返回当前 SqlSessionFactory 实例使用的 Configuration 对象。通过这个方法，可以获取到 MyBatis 的配置信息，并可以在运行时进行查询或修改。

6.3.2 SqlSession

SqlSession 在 MyBatis 中是非常强大的一个类，包含所有执行语句、提交或回滚事务以及获取映射器实例的方法。SqlSession 类的方法超过 20 个，常见方法如下：

1. 语句执行类方法

语句执行方法被用来执行定义在 SQL 映射文件中的 SELECT、INSERT、UPDATE 和 DELETE。其常见语句执行类方法中的参数 statement 代表映射语句的 ID，parameter 代表执行 SQL 语句的参数，具体如下：

(1) <T> T selectOne(String statement, Object parameter)：用于执行一条查询语句，并返回查询结果中的第一个记录，方法的返回类型是泛型 T，表示返回的对象类型。

(2) <E> List<E> selectList(String statement, Object parameter)：用于执行一条查询语句，并返回查询结果所有记录组成的列表。方法的返回类型是泛型 List<E>，表示返回的是列表元素类型。

(3) <T> Cursor<T> selectCursor(String statement, Object parameter)：用于执行

查询语句，但是返回的是一个 Cursor<T>对象，这个对象允许迭代读取查询结果集中的每一条记录，但它不会一次性将所有记录加载到内存中，实现数据的惰性加载。

(4) <K,V> Map<K,V> selectMap(String statement, Object parameter, String mapKey)：用于执行查询语句，并返回查询结果中的所有记录，但是以 Map 的形式返回，其中 mapKey 参数是指定将查询结果的哪个字段作为 Map 的键。

(5) int insert(String statement, Object parameter)：用于执行一条插入语句，将 parameter 对象映射的数据插入到数据库中，方法的返回值是当次操作影响的行数。

(6) int update(String statement, Object parameter)：用于执行一条更新语句，根据 parameter 对象映射的数据更新数据库中的记录，方法的返回值是当次操作影响的行数。

(7) int delete(String statement, Object parameter)：用于执行一条删除语句，根据 parameter 对象映射的条件删除数据库中的记录，方法的返回值是当次操作影响的行数。

2. 事务控制类方法

如果没有设置 MyBatis 自动提交，也没有使用外部事务管理器，则有以下 4 种方法来控制事务：

(1) void commit()：用于提交当前会话中执行的所有更改。

(2) void commit(boolean force)：与上面的 commit 方法类似，但它提供了一个可选的 force 参数。如果 force 设置为 true，即使没有任何更改，也会强制提交事务。

(3) void rollback()：用于撤销当前会话中执行的所有更改，任何在调用该方法之前已经提交到数据库的更改都将被撤销。

(4) void rollback(boolean force)；与 rollback 方法相似，如果 force 设置为 true，则会话中将进行强制回滚。

6.4　MyBatis 映射

MyBatis 映射是将 Java 接口的方法和 SQL 语句进行关联，或者将 Java 对象映射成数据库中的行记录，使开发者可以将更多的精力集中在业务逻辑上，而不是数据库操作的细节上，这极大地提高了开发效率和项目的可维护性。MyBatis 映射可以通过 XML 文件方式和注解方式两种主要方式实现。

6.4.1　XML 文件映射

MyBatis 的 XML 文件映射方式，是在 XML 文件中定义 SQL 语句、参数和结果映射，实现将 Java 对象映射成数据库中的记录。

1. <select>元素

在 XML 映射文件中<select>元素负责查询，其查询结果将被映射到 Java 对象。如下 XML 映射文件中返回一个或者多个 Student 类型的对象，对象属性值与查询

结果对应，具体如下：

```
<?xml version="1.0" encoding="UTF-8" ?>
<select id="selectOneStu" resultType="demo.Student" parameterType="Integer">
    select * from student where id = #{id}
</select>
<select id="selectAll" resultType="demo.Student">
    select * from student
</select>
```

第 1 个<select>标签定义一个名为 selectOneStu 的查询操作，它的作用是根据传入的整数参数 id 查询数据库中 student 表中的一项记录，并将其映射到 demo.Student 类型的对象中。resultType 属性指定映射的 Java 类型，parameterType 属性指定传入参数的类型；这里是一个整型，对应 SQL 查询语句 select * from student where id = #{id}中的 id 字段。第 2 个<select>标签定义一个名为 selectAll 的操作，它的作用是查询 student 表中的所有记录，并将每条记录映射到 demo.Student 类型的对象中。

注意：默认情况下，使用#{}参数语法时，MyBatis 会创建 PreparedStatement 参数占位符，并通过占位符安全地设置参数(就像使用 ? 一样)，而形如“${columnName}”参数代表直接插入一个不转义的字符串。

<select>元素属性如表 6-1 所示。

表 6-1 <select>元素属性

属 性	描 述
id	在命名空间中唯一的标识符，用于引用这条 SQL 语句
parameterType	传入参数的全限定类名或别名，由于 MyBatis 可通过类型处理器推断出具体传入语句的参数，这个属性非必选
resultType	从 SQL 语句中返回结果的全限定类名或别名，注意如果返回的是集合，应该设置为集合包含的类型，而不是集合本身的类型
resultMap	对外部 resultMap 的命名引用，详见 6.3.2 节，注意 resultType 和 resultMap 之间只能同时使用一个
flushCache	默认值为 false，将其设置为 true 后只要语句被调用，都会导致本地缓存和二级缓存被清空
useCache	默认值为 true，表示本条语句的结果被二级缓存缓存起来
timeout	表示等待数据库返回结果的超时时间(单位为秒)，默认值为 unset (表示未设置)
fetchSize	表示驱动程序每次批量返回的结果行数，默认值为 unset (表示未设置)
statementType	用于指定 SQL 语句的类型，取值有以下几种： · STATEMENT 类型，SQL 语句会被视为独立的语句，MyBatis 会执行该语句并返回结果 · PREPARED(默认)类型，语句会被预处理和存储，当再次执行相同的 SQL 语句时会重用，该类型可提供参数占位符 #{}用于动态绑定参数 · CALLABLE 类型，允许 MyBatis 执行带有输入参数和输出参数的存储过程

续表

属　性	描　　述
resultSets	这个设置仅适用于多结果集的情况，它将列出语句执行后返回的结果集并赋予每个结果集一个名称，多个名称之间以逗号分隔
databaseId	数据库 ID，如果在 MyBatis 配置文件中已配置数据库厂商标识，会加载匹配当前 databaseId 的语句

2. <insert>、<update>和<delete>元素

MyBatis Mapper 文件中，可以使用<insert>、<update>和<delete>元素定义插入、更新和删除 3 种数据库操作。示例代码如下：

```
<insert id="insertStu" parameterType="demo.Student">
        insert into student values(null, #{age}, #{name}, #{sex})
 </insert>
 <update id="updateData" parameterType="demo.Student">
       update student set age=#{age}, sex=#{sex}, name=#{name}   where id=#{id}
 </update>
 <delete id="deleteOne" parameterType="String">
      delete from student where name=#{name}
 </delete>
 <delete id="deleteAll">
     delete from student
 </delete>
```

以上代码中的具体解释如下：

<insert>元素定义一个插入操作，id 属性值为 insertStu，用于在 Java 代码中引用插入操作；parameterType 属性指定传入参数的类型，这里表示插入操作需要一个 demo.Student 类型的对象作为 SQL 插入语句 insert into student values(null, #{age}, #{name}, #{sex})的参数，而占位符#{age}、#{name}和#{sex}将由该对象中的属性值替换。

<update>元素定义一个更新操作，id 属性值为 updateData，用于在 Java 代码中引用更新操作；parameterType 属性指定传入参数的类型，这里为 demo.Student；SQL 语句表示更新 student 表中 age、sex 和 name 字段的值。

第 1 个<delete>元素定义一个删除操作，id 属性值为 deleteOne，用于在 Java 代码中引用删除操作。parameterType 属性指定传入参数的类型，这里表示删除操作需要一个 String 类型的对象作为参数。SQL 语句表示删除 student 表中 name 字段值为 #{name}的记录。

第 2 个<delete>元素定义一个删除所有记录的操作，其中 id 属性值为 deleteAll。这个操作不需要参数，因为 SQL 语句“delete from student”可以直接删除 student 表中的所有记录。

数据变更语句对应的<insert>、<update>和<delete>元素属性非常类似，如表

6-2 所示。

表 6-2 <insert>、<update>和<delete>元素属性

属　性	描　述
id	在命名空间中唯一的标识符，用于引用这条 SQL 语句
parameterType	传入参数的全限定类名或别名，由于 MyBatis 可通过类型处理器推断出具体传入语句的参数，这个属性不是必选
flushCache	默认为 true，表示只要语句被调用，都会导致本地缓存和二级缓存被清空
timeout	表示等待数据库返回结果的超时时间(秒)，默认值为 unset(代表未设置)
statementType	可选 STATEMENT、PREPARED 和 CALLABLE
useGeneratedKeys	仅适用于 insert 和 update，表示使用 getGeneratedKeys 方法取出由数据库内部生成的主键，默认值为 false
keyProperty	仅适用于 insert 和 update，指定主键值赋给 Java 对象的哪个属性，默认值为 unset(代表未设置)
keyColumn	仅适用于 insert 和 update，用于设置生成键值在表中的列名，某些数据库当主键列不是表中的第一列的时候，必须设置 keyColumn
databaseId	数据库 ID，如果在 MyBatis 配置文件中配置了数据库厂商标识，会加载匹配当前 databaseId 的语句

如果数据库支持自动生成主键，那么可以设置 useGeneratedKeys="true"，然后再把 keyProperty 设置为目标属性。例如，Author 表已经在主键 id 列上使用自动生成，那么语句可以修改为：

```
<insert id="insertAuthor" useGeneratedKeys="true"  keyProperty="id">
    insert into Author (username, password, email, bio)
    values (#{username}, #{password}, #{email}, #{bio})
</insert>
```

如果数据库还支持多行插入，也可以传入一个 Author 数组或集合，并返回自动生成的主键。相关代码如下：

```
<insert id="insertAuthor" useGeneratedKeys="true"  keyProperty="id">
    insert into Author (username, password, email, bio) values
    <foreach item="item" collection="list" separator=",">
        (#{item.username}, #{item.password}, #{item.email}, #{item.bio})
    </foreach>
</insert>
```

3. <resultMap>元素

之前已经讲过简单映射语句的示例，可以将所有的列映射到 resultType 属性指定的对象上，需要列名与对象属性名完全一致。当其不匹配的时候可以显式使用外部的<resultMap>元素方式，在引用它的语句中设置 resultMap 属性就行，同时去掉 resultType 属性。比如下面将表中的字段 user_id、user_name 和 hashed_password

映射到 User 的属性 id、username 和 password 中，具体代码如下：

```
<resultMap id="userResultMap" type="User">
    <id property="id" column="user_id" />
    <result property="username" column="user_name"/>
    <result property="password" column="hashed_password"/>
</resultMap>
<select id="selectUsers" resultMap="userResultMap">
    select user_id, user_name, hashed_password from some_table   where id = #{id}
</select>
```

以上这段代码中，<resultMap>元素定义如何从数据库结果集中映射到 Java 对象的一个元素，它的 id 属性定义映射的唯一标识符，type 属性指定要映射的 Java 类型是 User 类；<id>元素用来映射主键字段，property 属性指定 Java 对象中的属性名，而 column 属性指定数据库中对应的列名；<result>元素用来映射非主键字段，它的 property 属性和 column 属性分别代表 Java 对象的属性名和数据库的列名；<select>元素的 resultMap 属性引用前面定义的映射配置，告诉 MyBatis 使用这个映射来将查询结果映射到 User 对象。

4. <sql>和<include>元素

在 MyBatis 中，<sql>元素用于定义可重用的 SQL 片段，而<include>元素用于在其他的 SQL 映射语句中包含这些片段。这样可以提高 SQL 映射文件的可维护性，避免重复编写相同的 SQL 代码。例如：

```
<sql id="userColumns"> ${alias}.id, ${alias}.username, ${alias}.password </sql>
<select id="selectUsers" resultType="map">
    select
    <include refid="userColumns">
        <property name="alias" value="t1"/>
    </include>,
    <include refid="userColumns">
        <property name="alias" value="t2"/>
    </include>
        from some_table t1    cross join some_table t2
</select>
```

以上<sql>元素中定义了一个名为 userColumns 的 SQL 片段，包含了一个名为 alias 的属性，用于指定表别名；这个片段的作用是选择一个表中的 id、username 和 password 列。后面的<select>元素中使用<include>来包含 id 为 userColumns SQL 片段，使用 alias 属性将 t1 和 t2 将作为表别名替换到 SQL 片段中的 ${alias}占位符。

【案例 6-1】实现一个关于个人博客数据库管理的案例，具体实现步骤如下：

(1) 参照 6.1.2 节创建数据库脚本并运行，具体代码如下：

```
DROP DATABASE IF EXISTS test;
CREATE DATABASE test;
```

```
USE  test;
CREATE TABLE blog (
    id INT AUTO_INCREMENT PRIMARY KEY,
    title VARCHAR(255) NOT NULL,
    content TEXT NOT NULL,
    create_time TIMESTAMP DEFAULT CURRENT_TIMESTAMP
);
  INSERT INTO blog VALUES(NULL, 'java 心得体会 1', '学会了基本的条件语句...',
  CURRENT_TIMESTAMP);
  INSERT INTO blog VALUES(NULL, 'java 心得体会 2', '学会了基本的循环语句...',
  CURRENT_TIMESTAMP+1);
  INSERT INTO blog VALUES(NULL, 'java 心得体会 3', '学会了面向对象...',
  CURRENT_TIMESTAMP+2);
```

(2) 参照 6.1.2 节创建一个名称为 MybatisBasicconfig 的 Maven 项目并导入依赖。

(3) 创建 demo 包，在 demo 包下创建 Blog 类。

Blog.java

```
package demo;
import java.util.Date;
public class Blog {
    private Integer id;
    private String title;
    private String content;
    private Date create_time;
    public void setId(Integer id) {
        this.id = id;
    }
    public void setTitle(String title) {
        this.title = title;
    }
    public void setContent(String content) {
        this.content = content;
    }
    public void setcreate_time(Date create_time) {
        this.create_time = create_time;
    }
    @Override
    public String toString() {
        return "Blog{" +
                "id=" + id +
                ", title='" + title + '\'' +
```

```
                ", content='" + content + '\'' +
                ", create_time=" + create_time +
                '}';
    }
}
```

(4) 在 demo 包下创建 XML 映射文件 BasicMap.xml。具体代码如下：

```
<?xml version="1.0" encoding="UTF-8" ?>
<!DOCTYPE mapper
 PUBLIC "-//mybatis.org//DTD Mapper 3.0//EN"
 "http://mybatis.org/dtd/mybatis-3-mapper.dtd">
<mapper namespace="demo.BasicMap">
  <select id="selectOneBlog" resultType="demo.Blog" parameterType="Integer">
     select * from blog where id = #{id}
  </select>
  <select id="selectAll" resultType="demo.Blog">
       select * from blog
  </select>
  <insert id="insertBlog" parameterType="demo.Blog">
       insert into blog values(null, #{title}, #{content}, #{create_time})
  </insert>
  <update id="updateData" parameterType="demo.Blog">
      update blog set title=#{title}, content=#{content}, create_time=#{create_time}
      where id=#{id}
  </update>
  <delete id="deleteOne" parameterType="java.util.Date">
     delete from blog where create_time=#{create_time}
  </delete>
  <delete id="deleteAll">
    delete from blog
  </delete>
</mapper>
```

(5) 参照 6.1.2 节修改 mybatis-config.xml。

(6) 在 demo 包下创建测试类 Test.java。

Test.java

```
package demo;
import java.io.IOException;
import java.io.InputStream;
import java.util.Date;
import java.util.List;
import org.apache.ibatis.io.Resources;
```

```
import org.apache.ibatis.session.SqlSession;
import org.apache.ibatis.session.SqlSessionFactory;
import org.apache.ibatis.session.SqlSessionFactoryBuilder;
public class DemoTest {
    public static void main(String a[]) {
        try {
            InputStream inputStream = Resources.getResourceAsStream("mybatis-config.xml");
            SqlSessionFactory sessionFactory = new SqlSessionFactoryBuilder().
                build(inputStream);
            SqlSession session = sessionFactory.openSession();
            Blog blog = (Blog) session.selectOne("demo.BasicMap.selectOneBlog", 1);
            System.out.println(blog.toString());
            Date date=new Date();
            blog.setcreate_time(new Date());
            int rows = session.update("updateData", blog);
            System.out.println("更新:" + rows);
            rows = session.delete("deleteOne", date);
            System.out.println("删除:" + rows);
            List<Blog> blogs = session.selectList("selectAll");
            for (Blog b1 : blogs) {
                System.out.println(b1.toString());
            }
            rows = session.delete("deleteAll");
            System.out.println("删除:" + rows);
            blog.setId(1);
            blog.setTitle("学习 web 笔记");
            blog.setContent("掌握了 CSS...");
            rows = session.insert("insertBlog", blog);
            System.out.println("插入" + rows);
            blogs = session.selectList("selectAll");
            for (Blog b1 : blogs) {
                System.out.println(b1.toString());
            }
            session.commit();
            session.close();
        } catch (IOException e) {
            e.printStackTrace();
        }
    }
}
```

(7) 运行测试类 Test.java，结果如下：

```
Blog{id=1, title='java 心得体会 1', content='学会了基本的条件语句...', create_time=
Fri Apr 19 17:39:29 CST 2024}
更新:1
删除:1
Blog{id=2, title='java 心得体会 2', content='学会了基本的循环语句...', create_time=
Fri Apr 19 17:39:30 CST 2024}
Blog{id=3, title='java 心得体会 3', content='学会了面向对象...', create_time=
Fri Apr 19 17:39:31 CST 2024}
删除:2
插入 1
Blog{id=4, title='学习 web 笔记', content='掌握了 CSS...', create_time=
Fri Apr 19 17:41:28 CST 2024}
```

6.4.2　注解映射

注解映射是在 Java 代码中直接使用 SQL 语句和映射关系，这种方式使得 SQL 语句与代码更加紧密地结合在一起。

1. 映射器接口

@Mapper 注解用于标记接口为 MyBatis 的映射器接口。示例代码如下：

```
import org.apache.ibatis.annotations.Mapper;
@Mapper
public interface UserMapper {
   }
```

2. 增、删、改、查

增、删、改、查操作对应的注释如下：

(1) @Insert：插入操作，对应方法返回值表示影响的行数。

(2) @Update：更新操作，对应方法返回值表示影响的行数。

(3) @Delete：删除操作，对应方法返回值表示影响的行数。

(4) @Select：用于查询操作。

这几个注解等效于映射文件中<insert>、<delete>、<update>、<select>的功能，使用时只需要将 SQL 语句写在注解中。示例代码如下：

```
@Mapper
public interface BlogMapper {
    @Insert("INSERT INTO blog(title, content) VALUES(#{title}, #{content})")
    int insert(Blog blog);
    @Delete("DELETE FROM blog WHERE id = #{id}")
    int delete(Integer id);
    @Update("UPDATE blog SET title=#{title}, content=#{content}, updateTime=
    CURRENT_TIMESTAMP WHERE id = #{id}")
```

```
    int update(Blog blog);
    @Select("SELECT * FROM blog")
    List<Blog> selectAll();
    @Select("SELECT * FROM blog WHERE id = #{id}")
    Blog selectById(Integer id);
}
```

3. 映射参数

@Param 注解用于确保方法参数和映射文件中的 SQL 语句匹配，避免因参数名不一致而导致的错误。当方法有多个参数时，使用@Param 注解能够提高代码的可读性。示例代码如下：

```
@Select("SELECT * FROM users WHERE username = #{username} AND age = #{age}")
User getUserByUsernameAndAge(@Param("username") String uname, @Param("age") int age);
```

4. 映射结果

@Results 注解用于映射查询结果集，可以将查询结果映射到 Java 对象。示例代码如下：

```
@Results({
    @Result(property = "id", column = "user_id"),
    @Result(property = "name", column = "user_name"),
    @Result(property = "email", column = "user_email")
})
@Select("SELECT user_id, user_name, user_email FROM users")
List<User> getUsers();
```

【案例 6-2】注解方式实现 6.4.1 节个人博客数据库案例，具体实现步骤如下：

(1) 删除在 demo 包下的 XML 映射文件 BasicMap.xml。

(2) 在 demo 包下创建映射接口文件 BasicMap.java。

BasicMap.java

```
package demo;
import org.apache.ibatis.annotations.*;
import java.util.Date;
import java.util.List;
@Mapper
public interface BasicMap {
    @Select("select * from blog where id = #{id}")
    Blog selectOneBlog(@Param("id") Integer id);
    @Select("select * from blog")
    List<Blog> selectAll();
    @Insert("insert into blog values(null, #{title}, #{content}, #{create_time})")
    int insertBlog(Blog blog);
    @Update("update blog set title=#{title}, content=#{content},c reate_time=#{create_time}
    where id=#{id}")
```

```
    int updateData(Blog blog);
    @Delete("delete from blog where create_time=#{create_time}")
    int deleteOne(Date date);
    @Delete("delete from blog")
    int deleteAll();
}
```

(3) 修改 mybatis-config.xml 的 mapper 配置，相关代码如下：

```
<mappers>
    <mapper   class="demo.BasicMap"></mapper>
</mappers>
```

(4) 运行测试类 DemoTest，发现运行结果与 6.4.1 节的运行结果相同。

注意事项：对于稍微复杂一点的语句，Java 注解不仅力不从心，还会让本就复杂的 SQL 语句更加混乱不堪。因此如果需要做一些很复杂的操作，最好用 XML 来映射语句。

6.5　动态 SQL

在实际项目中需要根据不同条件拼接 SQL 语句，确保添加必要的空格，还要注意去掉多余的逗号。利用动态 SQL 可以彻底解决类似这些问题，常见语句如下：

1. <if>元素

使用动态 SQL 最常见的情景是根据条件包含 where 子句的一部分。示例代码如下：

```
<select id="findBlogWithTitleLike" resultType="demo.Blog">
        SELECT * FROM BLOG WHERE id>=#{id}
        <if test="title != null">
            AND title like #{title}
        </if>
  </select>
```

对应的 Java 代码如下：

```
Blog test=new Blog();
test.setId(1);
List<Blog> blogs = session.selectList("findBlogWithTitleLike", test);
System.out.println(blogs.size());
test.setTitle("java");
blogs = session.selectList("findBlogWithTitleLike", test);
System.out.println(blogs.size());
```

如果不传入“title”参数，所有满足 id 条件的 BLOG 都会返回；如果传入“title”参数，那么就会对“title”列进行模糊查找并返回对应的 BLOG 结果。

2. <choose>、<when>和<otherwise>元素

针对不想使用所有的条件，而只是想从多个条件中选择一个使用的情况，MyBatis 提供<choose>、<when>和<otherwise>元素，它有点像 Java 中的 switch 语句。例如，传入“title”就按“title”查找，传入“content”就按“content”查找；若两者都没有传入，就返回前 10 分钟的 BLOG。相关代码如下：

```
<select id="findBlogLike" resultType="demo.Blog">
        SELECT * FROM BLOG WHERE id>=#{id}
        <choose>
            <when test="title != null">
                AND title like #{title}
            </when>
            <when test="content != null">
                AND content like #{content}
            </when>
            <otherwise>
                AND create_time &lt;CURRENT_TIMESTAMP-600
            </otherwise>
        </choose>
    </select>
```

3. <trim>、<where>和<set>元素

<where>元素只会在子元素返回内容的情况下才插入“WHERE”子句；若子句的开头为“AND”或“OR”，<where>元素也会将它们去除。示例代码如下：

```
<select id="findBlogLikeA" resultType="demo.Blog">
        SELECT * FROM BLOG
        <where>
            <if test="title != null">
                title = #{title}
            </if>
            <if test="content != null">
                AND content like #{content}
            </if>
            <if test="create_time != null">
                AND create_time= #{create_time}
            </if>
        </where>
    </select>
```

与<where>元素等价的为自定义<trim>元素，实现条件插入 prefix 属性中指定的内容；其 prefixOverrides 属性表示若不满足条件将会忽略该文本序列。示例代码如下：

```
<select id="findBlogLikeB" resultType="demo.Blog">
        SELECT * FROM BLOG
```

```
        <trim prefix="WHERE" prefixOverrides="AND">
            <if test="title != null">
                title = #{title}
            </if>
            <if test="content != null">
                AND content like #{content}
            </if>
            <if test="create_time != null">
                AND create_time= #{create_time}
            </if>
        </trim>
</select>
```

<set>元素可用于动态包含需要更新的列，忽略其他不更新的列。例如：

```
<update id="updateBlog">
        UPDATE BLOG
        <set>
            <if test="title != null">title=#{title}, </if>
            <if test="content != null">content=#{content}, </if>
            <if test="create_time != null">create_time=#{create_time}, </if>
        </set>
        WHERE id=#{id}
 </update>>
```

以上例子中的<set>元素会动态地在行首插入 SET 关键字，并会删掉多余的逗号。

4. <foreach>元素

动态 SQL 的另一个常见使用场景是对集合进行遍历，尤其是在构建 IN 条件语句的时候，例如：

```
<select id="selectByIDs" resultType="demo.Blog">
        SELECT * FROM BLOG   WHERE id IN
        <foreach item="item" index="index" collection="list"
                 open="("   separator=", "   close=")">
            #{item}
        </foreach>
 </select>
```

以上例子中的<foreach>元素指定一个集合，声明在元素体内使用的集合项(item)和索引(index)变量。open、close 属性指定开头与结尾的字符串，separator 属性指定集合项迭代之间的分隔符。

5. <bind>元素

<bind>元素允许创建一个变量，并将其绑定到当前的上下文。例如：

```
<select id="findBlogByContent" parameterType="demo.Blog" resultType="demo.Blog">
        <bind name="pattern" value="'%'+_parameter.getContent()+'%'" />
```

```
        select * from blog   where   content like #{pattern}
    </select>
```

以上例子定义一个名为 pattern 的变量，这个变量将在 SQL 语句中用来作为 like 查询条件的模式。value 属性定义 pattern 变量的值是通过将_parameter 对象的 getContent()方法返回的值(即查询参数 demo.Blog 对象的 content 属性)与两个百分号%字符串拼接而成的。

6.6 实现 MyBatis 与 Spring 整合开发

MyBatis-Spring 是一个集成框架，它为 MyBatis 和 Spring 提供了无缝的集成。它允许 MyBatis 代码在 Spring 应用上下文中运行，并且能够利用 Spring 的强大功能，如依赖注入和事务管理。MyBatis-Spring 通过自动配置，简化了 Mapper 接口和 SqlSession 的创建过程，并将它们作为 Spring Bean 进行管理。

6.6.1 整合准备工作

实现 Spring 与 MyBatis 的整合开发，需要做一些必要的准备工作，具体如下：

(1) 在 4.1.2 节的 pom.xml 基础上增加以下依赖：mybatis-spring、mybatis、spring-jdbc、数据库驱动等。具体代码如下：

```
    <dependency>
            <groupId>org.springframework</groupId>
            <artifactId>spring-jdbc</artifactId>
            <version>5.3.24</version>
        </dependency>
        <dependency>
            <groupId>org.springframework</groupId>
            <artifactId>spring-tx</artifactId>
            <version>5.3.24</version>
        </dependency>
        <dependency>
            <groupId>org.mybatis</groupId>
            <artifactId>mybatis</artifactId>
            <version>3.5.10</version>
        </dependency>
        <dependency>
            <groupId>org.mybatis</groupId>
            <artifactId>mybatis-spring</artifactId>
            <version>2.0.7</version>
        </dependency>
        <!-- mysql 驱动包 -->
```

```
        <dependency>
            <groupId>mysql</groupId>
            <artifactId>mysql-connector-java</artifactId>
            <version>5.1.49</version>
        </dependency>
        <!--在 build 中配置 resources，防止资源导出失败的问题-->
       <build>
        <resources>
            <resource>
                <directory>src/main/resources</directory>
                <includes>
                    <include>**/*.properties</include>
                    <include>**/*.xml</include>
                </includes>
                <filtering>true</filtering>
            </resource>
            <resource>
                <directory>src/main/java</directory>
                <includes>
                    <include>**/*.properties</include>
                    <include>**/*.xml</include>
                </includes>
                <filtering>true</filtering>
            </resource>
        </resources>
    </build>
```

(2) 创建 MyBatis 配置文件 mybatis-config.xml，MyBatis 的配置文件中不再需要配置<dataSource>数据源信息，具体代码如下：

```
<?xml version="1.0" encoding="UTF-8" ?>
<!DOCTYPE configuration PUBLIC "-//mybatis.org//DTD Config 3.0//EN"
  "http://mybatis.org/dtd/mybatis-3-config.dtd">
  <configuration>
       <!--配置别名 -->
      <typeAliases>
           <package name="com.demo.po" />
       </typeAliases>
        <!--配置 Mapper 的位置 -->
        <mappers>
            <mapper resource="com/demo/mapper/xx.xml" />
         </mappers>
  </configuration>
```

以上配置中，typeAliases 元素用于定义类型别名，这样在 MyBatis 映射文件

中就可以使用类名的简称而不是全名。由于 SqlSessionFactoryBean 会创建它自有的 MyBatis 环境配置(Environment)，并按要求设置自定义环境的值，因此原 MyBatis 配置文件中<environments>、<DataSource>、<transactionManager>标签都会被忽略。

(3) Spring 的 XML 配置文件中增加以下 dataSource 和 sqlSessionFacory 配置，具体见 applicationContext.xml。

applicationContext.xml

```
<?xml version="1.0" encoding="UTF-8"?>
<beans xmlns="http://www.springframework.org/schema/beans"
      xmlns:xsi="http://www.w3.org/2001/XMLSchema-instance"
      xmlns:context="http://www.springframework.org/schema/context"
      xsi:schemaLocation="http://www.springframework.org/schema/beans
                  http://www.springframework.org/schema/beans/spring-beans.xsd
       http://www.springframework.org/schema/context
       http://www.springframework.org/schema/context/spring-context.xsd">
      <context:annotation-config></context:annotation-config>
   <!-- 定义数据源 -->
   <bean id="dataSource"
         class="org.springframework.jdbc.datasource.DriverManagerDataSource">
      <property name="driverClassName" value="com.mysql.jdbc.Driver"></property>
      <property name="url" value="jdbc:mysql://localhost:3306/test"></property>
      <property name="username" value="root"></property>
      <property name="password" value="root"></property>
   </bean>
    <!-- 配置 MyBatis 工厂 -->
   <bean id="sqlSessionFacory" class="org.mybatis.spring.SqlSessionFactoryBean">
      <!-- 注入数据源 -->
      <property name="dataSource" ref="dataSource"></property>
      <!-- 指定核心配置文件 -->
      <property name="configLocation" value="classpath:mybatis-config.xml"></property>
   </bean>
</beans>
```

以上代码定义了两个 Bean：其中 id 为“dataSource”的 Bean 对应 Spring 数据源实现类 DriverManagerDataSource，通过这个 Bean 的 JDBC 驱动类名、数据库连接 URL、用户名和密码的配置，实现了直接通过 JDBC 驱动管理数据库连接。另一个 id 为“sqlSessionFacory”的 Bean，用于创建 SqlSessionFactory 实例，其 class 属性指定了 MyBatis 工厂的实现类 SqlSessionFactoryBean，并通过<property>子元素注入数据源，指定了 MyBatis 的核心配置文件位置。

6.6.2 传统 DAO 方式整合

DAO 即 Data Access Object，是用户定义的用于访问数据库的对象，使用传统

DAO方式整合MyBatis和Spring，可通过SqlSessionTemplate和SqlSessionDaoSupport两种方式实现。

1. SqlSessionTemplate 方式

SqlSessionTemplate 是 SqlSession 的实现类，它的线程是安全的，它保证了使用 SqlSession 与当前 Spring 事务相关，可以被多个 DAO 或映射器所共享使用。使用该方式整合需要在Spring的配置文件中定义一个SqlSessionTemplate类型的Bean，并注入 SqlSessionFactory，然后在 DAO 实现类中注入这个 Bean。使用 SqlSession Template 类整合 MyBatis 与 Spring，具体实现步骤如下：

(1) 参考 6.1.2 节实现映射文件(如*mapper.XML)，并定义 SELECT、INSERT、UPDATE 和 DELETE 等 SQL 语句的映射；参考 6.1.2 节，修改 mybatis-config.xml 文件中关于映射文件路径的配置。

(2) 需要编写 DAO 接口和实现类。相关代码如下：

```
public interface StuDao {
    public Student findStuByID(Integer id);
    public Integer addStu(Student student);
    public List<Student> selectAll();
}
public class StuDaoImpl1   implements StuDao {
    @Autowired
    private SqlSession sqlSession;
    @Override
    public Student findStuByID(Integer id) {
        return sqlSession.selectOne("findStuByID", id);
    }
}
```

以上代码使用注解@Autowired 注入 sqlSession，findStuByID 方法中使用 SqlSession 的 selectOne 方法执行定义在映射文件中 id 为 findStuByID 的映射语句。

(3) 在 6.6.1 节 Spring 配置文件的基础上，增加创建 SqlSessionTemplate 和 DAO 接口的实现类 Bean，相关代码如下：

```
 <bean id="sqlSession" class="org.mybatis.spring.SqlSessionTemplate">
        <constructor-arg index="0" ref="sqlSessionFacory"></constructor-arg>
 </bean>
<bean id="stuDaoImpl1" class="TraditionDao.dao.StuDaoImpl1"></bean>
```

(4) 调用 DAO 接口操作数据库。相关代码如下：

```
@org.junit.Test
    public void testSqlSessionTemplate() {
        ApplicationContext context =
                        new ClassPathXmlApplicationContext("applicationContext.xml");
        StuDao stuDao=context.getBean("stuDaoImpl1", StuDao.class);
        Student student=stuDao.findStuByID(1);
```

```
        System.out.println(student.toString());
    }
```

2. SqlSessionDaoSupport 方式

SqlSessionDaoSupport 是 MyBatis 提供的一个抽象类，只需要 DAO 接口的实现类继承 SqlSessionDaoSupport 类，再注入一个 SqlSessionFactory，就可以直接使用 getSqlSession 方法来获取 SqlSession 实例。使用 SqlSessionDaoSupport 方式整合 MyBatis 和 Spring，具体步骤如下：

(1) 参考 6.1.2 节实现映射文件(如*mapper.XML)，并定义 SELECT、INSERT、UPDATE 和 DELETE 等 SQL 语句的映射；参考 6.1.2 节，修改 mybatis-config.xml 文件中关于映射文件路径的配置。

(2) 编写 DAO 接口和实现类 StuDao.java 和 StuDaoImpl2.java。

StuDao.java

```
public interface StuDao {
    public Student findStuByID(Integer id);
    public Integer addStu(Student student);
    public List<Student> selectAll();
}
```

StuDaoImpl2.java

```
public class StuDaoImpl2 extends SqlSessionDaoSupport implements StuDao {
    @Override
    public Student findStuByID(Integer id) {
        return getSqlSession().selectOne("findStuByID", id);
    }
}
```

由以上代码可以看出，实现类继承了 SqlSessionDaoSupport，可以在方法体内用 getSqlSession 获取 SqlSession，不用使用依赖注入的方式获取 SqlSession。

(3) 向实现类中注入 SqlSessionFactory。相关代码如下：

```
<bean id="stuDaoImpl2" class="TraditionDao.dao.StuDaoImpl2">
        <property name="sqlSessionFactory" ref="sqlSessionFacory"></property>
</bean>
```

(4) 调用 DAO 接口操作数据库。相关代码如下：

```
@org.junit.Test
    public void testDaoSupport() {
        ApplicationContext context =
                    new ClassPathXmlApplicationContext("applicationContext.xml");
        StuDao stuDao=context.getBean("stuDaoImpl2", StuDao.class);
        Student student=stuDao.findStuByID(1);
        System.out.println(student.toString());
}
```

6.6.3　MapperFactoryBean 方式整合

传统的 DAO 方式需要写 DAO 接口实现类，而实现类中要定义大量获取 SqlSession 的方法，并使用 SqlSession 执行在 SQL 映射文件(*mapper.XML)中的 SELECT、INSERT、UPDATE 和 DELETE 语句。而 MapperFactoryBean 是 MyBatis-Spring 团队提供的一个用于根据 Mapper 接口生成 Mapper 实现类，MapperFactoryBean 将会负责 SqlSession 的创建和关闭，用户只需要调用 Mapper 接口方法即可。

1. MapperFactoryBean 普通方式

使用 MapperFactoryBean 普通方式整合 MyBatis 和 Spring，需要在 Spring 配置文件中配置以下参数：

(1) mapperInterface：用于指定 Mapper 接口。

(2) SqlSessionFactory 或者 SqlSessionTemplate：指定获取 SqlSession 的 SqlSessionFactory 或者 SqlSessionTemplate。

使用 MapperFactoryBean 普通方式实现整合 MyBatis 和 Spring，具体实现步骤如下：

(1) 参考 6.1.2 节实现映射文件(如*mapper.XML)，并定义 SELECT、INSERT、UPDATE 和 DELETE 等 SQL 语句的映射；参考 6.1.2 节，修改 mybatis-config.xml 文件中关于映射文件路径的配置。

(2) 创建 Mapper 接口类 StuMapper.java，定义关于数据库操作的抽象方法。

StuMapper.java

```
package demoMapperDao.mapper;
import java.util.List;
import demoMapperDao.Po.Student;
public interface StuMapper {
    public Student findStuByID(Integer id);
}
```

(3) 通过 MapperFactoryBean 将接口加入 Spring 配置文件。相关代码如下：

```
<!-- MapperFactoryBean 方式，注意 map 文件和接口文件同名，且在同一个包下-->
<bean id="stuMapperNoImpl" class="org.mybatis.spring.mapper.MapperFactoryBean">
        <property name="mapperInterface" value="demoMapperDao.mapper.StuMapper">
        </property>
        <property name="sqlSessionFactory" ref="sqlSessionFacory"></property>
</bean>
```

以上代码声明一个 id 为 stuMapperNoImpl 的 MapperFactoryBean 类型的 Bean，定义了 mapperInterface 属性和 sqlSessionFactory 属性。

(4) 直接调用接口方法即可完成数据库操作。相关代码如下：

```
@org.junit.Test
        public void testMapperFactoryBean() {
```

```
            ApplicationContext context =
                        new ClassPathXmlApplicationContext ("applicationContext.xml");
            StuMapper stuMapper = context.getBean("stuMapperNoImpl", StuMapper.class);
            Student student = stuMapper.findStuByID(1);
            System.out.println(student.toString());
      }
```

从以上代码可以看出：不用编写 DAO 接口实现类，也不用获取 SqlSession，直接调用 Mapper 中的方法即可完成数据操作。

注意事项：

(1) 所指定的映射器类必须是一个接口，而不是具体的实现类。

(2) Mapper 接口的名称和对应的 Mapper.xml 映射文件的名称必须一致。

(3) Mapper.xml 文件中的 namespace 与 Mapper 接口的类路径相同。

(4) Mapper 接口中的方法名和 Mapper.xml 中定义的每个执行语句的 id 相同。

(5) Mapper 接口中方法的输入参数类型要和 Mapper.xml 中定义的每个 SQL 的 parameterType 类型相同。

(6) Mapper 接口方法的输出参数类型要和 Mapper.xml 中定义的每个 SQL 的 resultType 类型相同。

(7) 如果 SqlSessionTemplate 与 SqlSessionFactory 同时设定，则只会启用 SqlSessionTemplate。

2. MapperFactoryBean 注解方式

注解方式在 mapperFactoryBean 基础上去掉了映射文件，代码进一步减少，具体步骤如下：

(1) 创建工具接口类 StuDao，即 StuDao.java，添加注解@Select、@Insert 等。

StuDao.java

```
package demo.dao;
import java.util.List;
import demo.Po.Student;
import org.apache.ibatis.annotations.Insert;
import org.apache.ibatis.annotations.Param;
import org.apache.ibatis.annotations.Select;
public interface StuDao {
    @Select("SELECT * FROM student WHERE id = #{id}")
    public Student findStuByID(@Param("id") Integer id);
}
```

(2) 通过 MapperFactoryBean 将接口加入 Spring 配置文件，具体代码如下：

```
<bean id="stuDao" class="org.mybatis.spring.mapper.MapperFactoryBean">
        <property name="sqlSessionFactory" ref="sqlSessionFacory"/>
        <property name="mapperInterface" value="demo.dao.StuDao"/>
 </bean>
```

以上代码声明一个 id 为 stuMapperNoImpl 的 MapperFactoryBean 类型的 Bean，指定对应的 Mapper 接口为 StuMapper，指定对应 sqlSessionFactory。

(3) 直接调用接口方法，即可完成数据库操作。具体代码如下：

```
@org.junit.Test
    public void testBean() {
        ApplicationContext context =
                    new ClassPathXmlApplicationContext ("applicationContext.xml");
        StuDao stuDao=context.getBean("stuDao", StuDao.class);
        Student student=stuDao.findStuByID(1);
        System.out.println(student.toString());
}
```

可以看出注解方式不用编写映射文件和 Mapper 实现类，也不用获取 SqlSession，直接调用 Mapper 中的方法即可完成数据操作。

6.6.4　MapperScannerConfigurer 方式整合

上节介绍的 MapperFactoryBean 方式需要为每一个 Mapper 接口类声明一个 MapperFactoryBean 类型的 Bean，并指定对应 sqlSessionFactory。在实际的项目中，DAO 层会包含很多接口，如果每一个接口都在 Spring 配置文件中配置，不但会增加工作量，还会使得 Spring 配置文件非常臃肿。这时可以采用 MapperScannerConfigurer 类自动扫描的形式来配置映射器。

通常情况下，MapperScannerConfigurer 在使用时只需通过 basePackage 属性指定需要扫描的包即可，Spring 会自动通过包中的接口来生成映射器。在 Spring 配置文件中可以选择配置以下属性：

(1) basePackage：指定映射接口文件所在的包路径，当需要扫描多个包时可以使用分号或逗号作为分隔符。

(2) annotationClass：指定要扫描的注解名称，只有被注解标识的类才会被配置为映射器。

(3) sqlSessionFactoryBeanName：指定在 Spring 配置文件中定义的关于 SqlSession Factory 的 Bean 名称。

(4) sqlSessionTemplateBeanName：指定在 Spring 配置文件中定义的关于 Sql SessionTemplate 的 Bean 名称。

MapperScannerConfigurer 方式整合 MyBatis 和 Spring，具体实现步骤如下：

(1) 参考 6.1.2 节实现映射文件(如*mapper.XML)，并定义 SELECT、INSERT、UPDATE 和 DELETE 等 SQL 语句的映射。参考 6.1.2 节，修改 mybatis-config.xml 文件中关于映射文件路径的配置。

(2) 需先定义一个或者多个 Mapper 接口，例如 StuDao .java。

StuDao.java

```
package demoMapperScanner.dao;
import java.util.List;
```

```
import demoMapperScanner.Po.Student;
public interface StuDao {
    public Student findStuByID(Integer id);
}
```

(3) 通过将 MapperScannerConfigurer 接口加入 Spring 配置文件，配置对应的扫描路径，再注入 sqlSessionFacory，具体格式如下：

```
<bean class="org.mybatis.spring.mapper.MapperScannerConfigurer">
        <property name="basePackage" value="demoMapperScanner"/>
        <property name="sqlSessionFactoryBeanName" value="sqlSessionFacory"/>
 </bean>
```

(4) 直接调用接口方法，即可完成数据库操作，具体格式如下：

```
  @org.junit.Test
        public void testMapperFactoryBean() {
            ApplicationContext context =
                        new ClassPathXmlApplicationContext("applicationContext.xml");
            StuDao stuDao=context.getBean("stuDao", StuDao.class);
            Student student=stuDao.findStuByID(1);
            System.out.println(student.toString());
}
```

6.7 实战演练：驾校学员系统数据访问层实现

1. 需求分析

需求分析如下：

序号	需求类别	需 求 描 述
1	功能需求	添加学员：允许用户输入学员信息并将其添加到学员表中
		查询学员：允许用户根据 id、姓名、年龄、性别等条件查询学员信息
		修改学员：允许用户修改已存在的学员信息
		删除学员：允许用户通过 id 删除指定学员
		查询所有学员：显示所有学员的信息
2	数据访问层需求	学员 Mapper 接口：定义学员表的操作接口，包括添加、查询、修改、删除等操作
		学员 Mapper XML 文件：编写对应的 SQL 语句，实现接口中的方法
		数据库表结构：设计学员表，包括 id、姓名、年龄、性别等字段

2. 规划设计

填写如下规划设计表：

WBS 表			
项目基本情况			
项目名称	驾校学员信息管理系统数据访问层	任务编号	
姓　　名		班　　级	
工作分解			
工作任务	包含活动	备注	
1. 数据库设计	1.1　创建学员表		
	1.2　测试数据库		
2. 创建项目基本结构	2.1　创建 Maven 工程		
	2.2　创建实体类		
	2.2　数据库、MyBatis、Spring 的基本配置		
	2.3　测试项目基本结构		
3. 核心功能实现	3.1　学员信息录入、删除、修改、查询功能		
	3.2　测试数据访问层		

3. 项目阶段一：搭建环境

(1) 创建数据库脚本 data.sql，用于创建表和记录。

data.sql

```
DROP DATABASE IF EXISTS test;
CREATE DATABASE test;
USE   test;
CREATE TABLE student(
    id   INT   NOT NULL AUTO_INCREMENT,
    age INT,
    name VARCHAR(20),
    gender VARCHAR(20),
    PRIMARY KEY(id)
    );
  INSERT INTO student VALUES(NULL, 32, '李明', '男');
  INSERT INTO student VALUES(NULL, 33, '王伟', '女');
  INSERT INTO student VALUES(NULL, 34, '孙明', '男');
  INSERT INTO student VALUES(NULL, 31, '王伟大', '女');
```

以上脚本可实现创建名称为 test 的数据库和 student 表，最后插入 4 条记录。

(2) 输入以下指令：

执行数据库脚本（data.sql），结果如图 6-5 所示。

```
mysql –u 密码 –p
source data.sql
```

```
mysql> source data.sql
Query OK, 1 row affected (0.00 sec)

Query OK, 1 row affected (0.00 sec)

Database changed
Query OK, 0 rows affected (0.00 sec)

Query OK, 1 row affected (0.02 sec)

Query OK, 1 row affected (0.00 sec)

Query OK, 1 row affected (0.00 sec)

Query OK, 1 row affected (0.00 sec)
```

图 6-5 导入数据库脚本结果

4. 项目阶段二：项目基本结构创建

(1) 参考 4.1.2 节创建 Maven 项目，在 pom.xml 中添加 spring、mybatis 和 mysql 等相关依赖。

pom.xml

```xml
<?xml version="1.0" encoding="UTF-8"?>
<project xmlns="http://maven.apache.org/POM/4.0.0"
         xmlns:xsi="http://www.w3.org/2001/XMLSchema-instance"
         xsi:schemaLocation="http://maven.apache.org/POM/4.0.0
                             http://maven.apache.org/xsd/maven-4.0.0.xsd">
    <modelVersion>4.0.0</modelVersion>
    <groupId>org.example</groupId>
    <artifactId>MybatisSpringTranditionalDao</artifactId>
    <version>1.0-SNAPSHOT</version>
    <properties>
        <maven.compiler.source>8</maven.compiler.source>
        <maven.compiler.target>8</maven.compiler.target>
    </properties>
    <dependencies>
        <dependency>
            <groupId>org.springframework</groupId>
            <artifactId>spring-core</artifactId>
            <version>5.3.24</version>
        </dependency>
        <dependency>
            <groupId>org.springframework</groupId>
            <artifactId>spring-beans</artifactId>
            <version>5.3.24</version>
        </dependency>
        <dependency>
            <groupId>org.springframework</groupId>
```

```
        <artifactId>spring-context</artifactId>
        <version>5.3.24</version>
    </dependency>
    <dependency>
        <groupId>org.springframework</groupId>
        <artifactId>spring-aop</artifactId>
        <version>5.3.24</version>
    </dependency>
    <dependency>
        <groupId>org.springframework</groupId>
        <artifactId>spring-expression</artifactId>
        <version>5.3.24</version>
    </dependency>
    <dependency>
        <groupId>org.springframework</groupId>
        <artifactId>spring-jcl</artifactId>
        <version>5.3.24</version>
    </dependency>
    <dependency>
        <groupId>org.springframework</groupId>
        <artifactId>spring-jdbc</artifactId>
        <version>5.3.24</version>
    </dependency>
    <dependency>
        <groupId>org.springframework</groupId>
        <artifactId>spring-tx</artifactId>
        <version>5.3.24</version>
    </dependency>
    <dependency>
        <groupId>org.mybatis</groupId>
        <artifactId>mybatis</artifactId>
        <version>3.5.10</version>
    </dependency>
    <dependency>
        <groupId>org.mybatis</groupId>
        <artifactId>mybatis-spring</artifactId>
        <version>2.0.7</version>
    </dependency>
    <!-- mysql 驱动包 -->
    <dependency>
```

```
            <groupId>mysql</groupId>
            <artifactId>mysql-connector-java</artifactId>
            <version>5.1.49</version>
        </dependency>
        <dependency>
            <groupId>junit</groupId>
            <artifactId>junit</artifactId>
            <version>4.8.2</version>
        </dependency>
</dependencies>
<!--在 build 中配置 resources，来防止资源导出失败的问题-->
    <build>
        <resources>
            <resource>
                <directory>src/main/resources</directory>
                <includes>
                    <include>**/*.properties</include>
                    <include>**/*.xml</include>
                </includes>
                <filtering>true</filtering>
            </resource>
            <resource>
                <directory>src/main/java</directory>
                <includes>
                    <include>**/*.properties</include>
                    <include>**/*.xml</include>
                </includes>
                <filtering>true</filtering>
            </resource>
        </resources>
    </build>
</project>
```

(2) 在Resources目录下创建属性文件db.properties，注意内容必须是jdbc.***。

db.properties

```
jdbc.driver =com.mysql.jdbc.Driver
jdbc.url =jdbc:mysql://localhost:3306/test
jdbc.username =root
jdbc.password =root
```

以上配置中 jdbc.driver 指定了要使用的 JDBC 驱动程序；jdbc.url 指定了数据库的 URL，包括数据库类型(mysql)、主机地址(localhost)、端口号(3306)和数据库

名(test)；jdbc.username 指定了连接数据库时使用的用户名；jdbc.password 指定了连接数据库时使用的密码。

(3) 在 Resources 目录下创建 MyBatis 的核心配置文件 mybatis-config.xml，从官网拷贝模板后修改。

mybatis-config.xml

```
<?xml version="1.0" encoding="UTF-8" ?>
<!DOCTYPE configuration
    PUBLIC "-//mybatis.org//DTD Config 3.0//EN"
    "http://mybatis.org/dtd/mybatis-3-config.dtd">
<configuration>
<settings>
        <setting name="logImpl" value="STDOUT_LOGGING"/>
</settings>
<typeAliases>
      <package name="demoStudentInfo.Po" />
</typeAliases>
</configuration>
```

以上配置完成了将日志输出到控制台，并将包名为“demoStudentInfo.Po”下的所有类都注册为别名，别名就是类的简单名称(不包含包名)。

(4) 在创建 Spring 配置文件 applicationContext.xml，完成配置和管理 Spring 应用程序中的 Bean，包括数据源、事务管理器和 MyBatis 等组件。

applicationContext.xml

```
<?xml version="1.0" encoding="UTF-8"?>
<beans xmlns="http://www.springframework.org/schema/beans"
       xmlns:xsi="http://www.w3.org/2001/XMLSchema-instance"
       xmlns:aop="http://www.springframework.org/schema/aop"
       xmlns:context="http://www.springframework.org/schema/context"
       xmlns:tx="http://www.springframework.org/schema/tx"
       xsi:schemaLocation="
         http://www.springframework.org/schema/beans
         https://www.springframework.org/schema/beans/spring-beans.xsd
         http://www.springframework.org/schema/aop
         https://www.springframework.org/schema/aop/spring-aop.xsd
         http://www.springframework.org/schema/context
         https://www.springframework.org/schema/context/spring-context.xsd
         http://www.springframework.org/schema/tx
         https://www.springframework.org/schema/tx/spring-tx.xsd">
    <!-- 读取属性文件 -->
    <context:property-placeholder location="classpath:db.properties" />
    <!-- 添加注解驱动 -->
    <context:annotation-config></context:annotation-config>
```

```
        <!-- 定义数据源 -->
        <bean id="dataSource"
            class="org.springframework.jdbc.datasource.DriverManagerDataSource">
            <property name="driverClassName" value="${jdbc.driver}"></property>
            <property name="url" value="${jdbc.url}"></property>
            <property name="username" value="${jdbc.username}"></property>
            <property name="password" value="${jdbc.password}"></property>
        </bean>
        <!-- 定义事务管理器 -->
        <bean id="transactionManager" class="org.springframework.jdbc.datasource.
        DataSourceTransactionManager">
            <property name="dataSource" ref="dataSource"></property>
        <!-- 指定核心配置文件 -->
            <property name="configLocation" value="classpath:mybatis-config.xml"></property>
        </bean>
        <!-- 开启事务注解 -->
        <tx:annotation-driven transaction-manager="transactionManager" />
        <!-- 配置 MyBatis 工厂 -->
        <bean id="sqlSessionFacory" class="org.mybatis.spring.SqlSessionFactoryBean">
            <!-- 注入数据源 -->
            <property name="dataSource" ref="dataSource"></property>
            <!-- 指定核心配置文件 -->
            <property name="configLocation" value="classpath:mybatis-config.xml"></property>
        </bean>
        <bean id="student" class=" demoStudentInfo.Po.Student">
            <property name="age" value="15"/>
            <property name="id" value="001"/>
            <property name="name" value="wang dong"/>
            <property name="gender" value="boy"/>
        </bean>
    </beans>
```

(5) 创建 demoStudentInfo 包，该包下面创建 Po 包和 mapper 包，这两个包下分别创建实体类 Student 和工具接口类 StuMapper，即 Student.java 和 StuMapper.java。

Student.java

```
package demoStudentInfo.Po;
public class Student {
    public int getId() {
    return id;
    }
    public void setId(int id) {
        this.id = id;
```

```
    }
    public int getAge() {
        return age;
    }
    public void setAge(int age) {
        this.age = age;
    }
    public String getName() {
        return name;
    }
    public void setName(String name) {
        this.name = name;
    }
    public String getGender() {
        return gender;
    }
    public void setGender(String gender) {
        this.gender = gender;
    }
    @Override
    public String toString() {
        return "Student [id=" + id + ", age=" + age + ", name=" + name + ", gender=" + gender + "]";
    }
    int id;
    int age;
    String name;
    String gender;
}
```

以上实体类 Student 代表数据库中的 student 表，它包含了与表中的列对应的属性。

StuMapper.java

```
package demoStudentInfo.mapper;
import java.util.List;
import demoStudentInfo.Po.Student;
import org.springframework.transaction.annotation.Transactional;
@Transactional
public interface StuMapper {
    public Student findStuByID(Integer id);
    public Integer addStu(Student student);
    public List<Student> selectAll();
}
```

以上工具接口类 StuMapper 用于定义与数据库交互的 SQL 映射语句。

(6) 在 demoStudentInfo.mapper 包下创建测试类 Test，即 Test.java，用于对系统的初步测试。

Test.java

```
package demoStudentInfo.mapper;
import org.springframework.context.ApplicationContext;
import org.springframework.context.support.ClassPathXmlApplicationContext;
import demoStudentInfo.Po.Student;
public class Test {
    @org.junit.Test
    // Spring 环境测试
    public void springTest() {
        ApplicationContext context =
                    new ClassPathXmlApplicationContext ("applicationContext.xml");
        Student student = context.getBean("student", Student.class);
        System.out.println(student.toString());
    }
}
```

(7) Spring 基本环境测试：选中 springTest 方法，单击右键运行，结果如下：

```
Student [id=1, age=15, name=wang dong, gender=boy]
Process finished with exit code 0
```

由以上结果看出，Spring 完成了 Student 实例的创建，说明 Spring 基本环境正常。

5. 项目阶段三：核心功能方式实现

(1) 在 demoStudentInfo.mapper 下创建映射文件 StuMapper.xml，定义 SQL 语句和映射信息。

StuMapper.xml

```
<?xml version="1.0" encoding="UTF-8" ?>
<!DOCTYPE mapper
        PUBLIC "-//mybatis.org//DTD Mapper 3.0//EN"
        "http://mybatis.org/dtd/mybatis-3-mapper.dtd">
<mapper namespace="demoStudentInfo.mapper.StuMapper">
    <!-- MyBatis 配置文件中已经指定包名，会在包名下面搜索需要的 Java Bean -->
    <select id="findStuByID" resultType="Student">
        select * from student where id = #{id}
    </select>
    <insert id="addStu" parameterType="Student">
        insert into student values(null, #{age}, #{name}, #{sex})
    </insert>
```

```
    <select id="selectAll" resultType="Student">
        select * from student
    </select>
</mapper>
```

以上代码定义了一个查询操作，用于根据学生 ID 查找学生记录；定义了一个插入操作，用于向学生表中添加一条新的学生记录；定义了一个查询操作，用于检索学生表中的所有记录。

(2) 修改 MyBatis 的配置文件 mybatis-config.xml，配置 mapper 位置，即增加以下配置：

```
<mappers>
    <mapper resource="demoStudentInfo/mapper/StuMapper.xml" />
</mappers>
```

(3) 创建和修改 Spring 配置文件 applicationContext.xml，添加关于 Mapper FactoryBean 的配置，即增加以下 Bean：

```
<bean id="stuMapperNoImpl" class="org.mybatis.spring.mapper.MapperFactoryBean">
        <property name="mapperInterface" value="demoStudentInfo.mapper.StuMapper">
        </property>
        <property name="sqlSessionFactory" ref="sqlSessionFacory"></property>
</bean>
```

(4) 在阶段二所创建的 Test.java 中，使用 Spring 和 MyBatis 进行数据库的查询、更新和插入操作，即增加以下方法：

```
@org.junit.Test
    public void testDaoSupport() {
        ApplicationContext context =
                        new ClassPathXmlApplicationContext ("applicationContext.xml");
        StuMapper stuMapper = context.getBean("stuMapperNoImpl", StuMapper.class);
        Student student = stuMapper.findStuByID(1);
        System.out.println(student.toString());
        stuMapper.selectAll();
        student.setAge(34);
        student.setName("小王");
        stuMapper.addStu(student);
        stuMapper.selectAll();
}
```

以上代码从 Spring 容器中获取名为 stuMapperNoImpl 的 Bean，是 StuMapper 接口的一个代理实现，通过这个代理对象来执行 SQL 映射语句完成数据库操作。

(5) 右键单击 testDaoSupport 方法运行，结果如下：

```
==>  Preparing: select * from student where id = ?
==> Parameters: 1(Integer)
<==     Columns: id, age, name, gender
```

```
<==        Row: 1, 32, 李明, 男
<==      Total: 1
Releasing transactional SqlSession
Student [id=1, age=32, name=李明, gender=男]
Creating a new SqlSession
Registering transaction synchronization for SqlSession
==>  Preparing: select * from student
==> Parameters:
<==    Columns: id, age, name, gender
<==        Row: 1, 32, 李明, 男
<==        Row: 2, 33, 王伟, 女
<==        Row: 3, 34, 孙明, 男
<==        Row: 4, 31, 王伟大, 女
<==      Total: 4
Releasing transactional SqlSession
Student [id=1, age=32, name=李明, gender=男]
Student [id=2, age=33, name=王伟, gender=女]
Student [id=3, age=34, name=孙明, gender=男]
Student [id=4, age=31, name=王伟大, gender=女]
Transaction synchronization committing SqlSession
Creating a new SqlSession
Registering transaction synchronization for SqlSession
==>  Preparing: insert into student values(null, ?, ?, ?)
==> Parameters: 34(Integer), 小王(String), 男(String)
<==    Updates: 1
Releasing transactional SqlSession
Creating a new SqlSession
Registering transaction synchronization for SqlSession
==>  Preparing: select * from student
==> Parameters:
<==    Columns: id, age, name, gender
<==        Row: 1, 32, 李明, 男
<==        Row: 2, 33, 王伟, 女
<==        Row: 3, 34, 孙明, 男
<==        Row: 4, 31, 王伟大, 女
<==        Row: 7, 34, 小王, 男
<==      Total: 5
Releasing transactional SqlSession
Student [id=1, age=32, name=李明, gender=男]
Student [id=2, age=33, name=王伟, gender=女]
Student [id=3, age=34, name=孙明, gender=男]
Student [id=4, age=31, name=王伟大, gender=女]
```

```
Student [id=7, age=34, name=小王, gender=男]
```

可以看出在整合 Spring 和 MyBatis 后进行数据库操作，查询、更新和插入操作结果均正确。

强化练习

习题答案

1. MyBatis 是(　　)。

A. 数据库连接池　　B. 持久层框架

C. 动态 SQL 生成器　　D. 数据库访问层

2. (　　)是 MyBatis 映射文件中的常见标签。

A. select　　B. insert　　C. update　　D. delete

3. 在 MyBatis 中，通过(　　)实现插入数据。

A. 使用 INSERT 标签　　B. 使用 SELECT 标签

C. 使用 UPDATE 标签　　D. 使用 DELETE 标签

4. (　　)关键字用于实现 MyBatis 查询数据。

A. SELECT　　B. INSERT　　C. UPDATE　　D. DELETE

5. MyBatis 中结果映射(ResultMap)的作用是(　　)。

A. 将 SQL 查询结果映射到 Java 对象

B. 实现 Java 对象与数据库表字段的映射

C. 定义查询语句

D. 定义事务管理策略

6. (　　)注解用于 MyBatis 事务管理。

A. @Transactional　　B. @Before

C. @After　　D. @Override

7. MyBatis 中的映射文件有(　　)类型。

A. Mapper 映射文件　　B. Mapper 接口

C. Service 接口　　D. Controller 类

进一步学习建议

学习 MyBatis 之后，可以进一步学习以下内容，以提高技术、技能：

(1) 数据库优化：学习优化数据库查询语句的方法，提高应用程序的性能。了解 SQL 语句的执行计划，分析慢查询，使用索引、分页等技巧优化查询速度。

(2) 探索国产数据库的使用：MyBatis 是一款优秀的持久层框架，可以兼容多种数据库，包括国产数据库。MyBatis 支持 JDBC 协议，只要国产数据库提供 JDBC 驱动，就可以在项目中使用 MyBatis 进行数据库操作。

考核评价

<table>
<tr><td colspan="4">考核评价表</td></tr>
<tr><td colspan="2">姓名</td><td></td><td>班　　级</td><td></td></tr>
<tr><td colspan="2">学号</td><td></td><td>考评时间</td><td></td></tr>
<tr><td colspan="2">评价主题及总分</td><td>评价内容及分数</td><td>评分</td></tr>
<tr><td rowspan="3">1</td><td rowspan="3">知识考核
(30)</td><td>阐述 MyBatis 的核心配置文件结构(10 分)</td><td></td></tr>
<tr><td>举例说明 MyBatis 的映射文件常用元素的作用(10 分)</td><td></td></tr>
<tr><td>MyBatis 的注解方式有什么优缺点(10 分)</td><td></td></tr>
<tr><td rowspan="3">2</td><td rowspan="3">技能考核
(40)</td><td>具备业务需求分析、功能设计、编码及测试的综合能力(10 分)</td><td></td></tr>
<tr><td>开发任务能够按时完成(20 分)</td><td></td></tr>
<tr><td>熟练使用 MyBatis 进行数据库操作(增删改查)(10 分)</td><td></td></tr>
<tr><td rowspan="3">3</td><td rowspan="3">思政考核
(30)</td><td>列举我国在数据库技术领域的发展历程和取得的成就(10 分)</td><td rowspan="3"></td></tr>
<tr><td>具备良好的团队协作精神，能与团队成员有效沟通(10 分)</td></tr>
<tr><td>在解决问题时，能总结经验教训，避免类似问题再次出现(10 分)</td></tr>
<tr><td colspan="3">评语：</td><td>汇总：</td></tr>
</table>

第 7 章　综合项目——智慧农业数据分析平台

学习目标

目标类型	目 标 描 述
知识目标	• 巩固 ECharts 常用图表的原理、基本参数配置 • 理解 ECharts 的交互功能及扩展功能的实现过程 • 掌握 Spring 框架的基本配置、Spring MVC 的工作流程、MyBatis 的核心配置及 SSM 框架的技术原理 • 掌握 WebSocket 通信原理及实现过程
技能目标	• 掌握 ECharts 常用基本组件和图表的绘制方法，能根据不同场景需求选择不同类型图表进行数据展示 • 能够使用 SSM 框架技术搭建项目，并结合 ECharts 组件完成不同场景的业务需求
思政目标	• 提高团队合作开发和沟通能力，培养集体意识和协作精神，增强团队成员之间信任感，提高解决冲突的能力 • 培养独立解决问题的能力，勇敢地直面各种困难和挑战 • 培养编程思维、实践能力以及解决复杂问题的能力，理论和实践相结合才能将知识转化得更好、更快 • 培养职业道德和社会责任感，通过专业技术改善农业发展过程中的问题，助力乡村振兴

项目实战

7.1　项 目 概 述

随着物联网和大数据技术的发展，农业生产正逐渐向智能化、精准化转变。为了提升农业生产效率，优化资源配置，本项目将开发一个农业数据分析平台，该平台能够实现农业生产相关数据的实时采集、存储、处理和可视化展示，帮助企业及时了解各地区的农业生产状况。项目的功能主要包括：

(1) 数据采集：收集农业生产相关数据，如气温、光照、风速、降雨量和农作物产量等。

(2) 数据存储：将收集到的数据存储在 MySQL 数据库中，以便进行后续分析。

(3) 数据处理：使用 SSM 框架搭建数据处理层，对原始数据进行清洗、转换和整合，生成可供前端展示的图表数据。

(4) 数据可视化：使用 ECharts 绘制各类图表，如仪表盘图、折线图、柱状图和环形图等实时展示农业生产数据，使得企业能够清晰地了解各地区的农业生产状况，为后续的决策提供有力支持。

7.2 项 目 规 划

填写如下项目规划设计表：

<table>
<tr><td colspan="4">WBS 表</td></tr>
<tr><td colspan="4">项目基本情况</td></tr>
<tr><td>项目名称</td><td>智慧农业数据分析平台</td><td>任务编号</td><td></td></tr>
<tr><td>姓　　名</td><td></td><td>班　　级</td><td></td></tr>
<tr><td colspan="4">工作分解</td></tr>
<tr><td>工作任务</td><td colspan="2">包含活动</td><td>备注</td></tr>
<tr><td rowspan="2">1. 数据库设计</td><td colspan="2">1.1 表 weather 的设计与创建</td><td></td></tr>
<tr><td colspan="2">1.2 表 yield 的设计与创建</td><td></td></tr>
<tr><td rowspan="3">2. 系统环境搭建</td><td colspan="2">2.1 创建项目</td><td></td></tr>
<tr><td colspan="2">2.2 添加项目依赖</td><td></td></tr>
<tr><td colspan="2">2.3 添加项目相关包</td><td></td></tr>
<tr><td rowspan="4">3. 数据采集功能</td><td colspan="2">3.1 收集某地区光照、温度和降雨量等数据</td><td></td></tr>
<tr><td colspan="2">3.2 创建 Mapper 文件</td><td></td></tr>
<tr><td colspan="2">3.3 创建 DAO 类</td><td></td></tr>
<tr><td colspan="2">3.4 创建 POJO 类</td><td></td></tr>
<tr><td rowspan="2">4. 数据存储功能</td><td colspan="2">4.1 设计数据存储逻辑</td><td></td></tr>
<tr><td colspan="2">4.2 设计数据处理逻辑</td><td></td></tr>
<tr><td rowspan="3">5. 数据处理功能</td><td colspan="2">5.1 创建控制器类</td><td></td></tr>
<tr><td colspan="2">5.2 创建通信类</td><td></td></tr>
<tr><td colspan="2">5.3 创建 Spring 配置文件</td><td></td></tr>
<tr><td rowspan="9">6. 数据可视化</td><td colspan="2">6.1 添加前端资源如 js 库</td><td></td></tr>
<tr><td colspan="2">6.2 创建 jsp 页面文件</td><td></td></tr>
<tr><td colspan="2">6.3 实现温度数据可视化</td><td></td></tr>
<tr><td colspan="2">6.4 实现光照数据可视化</td><td></td></tr>
<tr><td colspan="2">6.5 实现风速数据可视化</td><td></td></tr>
<tr><td colspan="2">6.6 实现降雨量数据可视化</td><td></td></tr>
<tr><td colspan="2">6.7 实现农作物产量数据可视化</td><td></td></tr>
<tr><td colspan="2">6.8 配置 Tomcat 运行项目</td><td></td></tr>
<tr><td colspan="2">6.9 测试</td><td></td></tr>
</table>

7.3　数据库设计

7.3.1　创建 weather 表

新建 weather.sql 脚本文件，用于创建 weather 表，该表用于存储气温、光照、降水量等农业生产相关数据。脚本文件 weather.sql 中 id 的字段为数据库主键，date 字段为日期，region 字段为地区，sunshine 字段为光照时长，avgwind 字段为平均风速，maxtemp 字段为最高温度，mintemp 字段为最低温度，avgtemp 字段为平均温度，envhumid 字段为湿度，avgpress 字段为平均气压，precipitation 字段为降雨量。

weather.sql

```
DROP TABLE IF EXISTS 'weather';
CREATE TABLE 'weather' (
    'id' int(11) NOT NULL AUTO_INCREMENT,
    'date' varchar(20) DEFAULT '',
    'region' varchar(255) CHARACTER SET gbk COLLATE gbk_chinese_ci DEFAULT '',
    'sunshine' int(11) DEFAULT NULL,
    'avgwind' double(11, 1) DEFAULT NULL,
    'maxtemp' double(11, 1) DEFAULT NULL,
    'mintemp' double(11, 1) DEFAULT NULL,
    'avgtemp' double(11, 1) DEFAULT NULL,
    'envhumid' double(11, 1) DEFAULT NULL,
    'avgpress' int(11) DEFAULT NULL,
    'precipitation' int(11) DEFAULT NULL,
    PRIMARY KEY ('id')
) ENGINE=InnoDB AUTO_INCREMENT=374 DEFAULT CHARSET=latin1 COLLATE=
latin1_swedish_ci;
```

7.3.2　创建 yield 表

新建 yield.sql 脚本文件，用于创建 yield 表，该表用于存储各地区农作物产量和种植面积等数据。脚本文件 yield.sql 中的 id 字段为数据库主键，year 字段为年份，region 字段为地区，crop 字段为农作物类型，yield 字段为产量，area 字段为种植面积。

yield.sql

```
DROP TABLE IF EXISTS 'yield';
CREATE TABLE 'yield' (
    'id' int(11) NOT NULL AUTO_INCREMENT,
```

```
    'year' varchar(255) DEFAULT NULL,
    'region' varchar(255) CHARACTER SET gbk COLLATE gbk_chinese_ci DEFAULT '',
    'crop' varchar(255) CHARACTER SET gbk COLLATE gbk_chinese_ci DEFAULT '',
    'yield' double(11, 2) DEFAULT NULL,
    'area' double(11, 2) DEFAULT NULL,
    PRIMARY KEY ('id')
) ENGINE=InnoDB AUTO_INCREMENT=11 DEFAULT CHARSET=latin1 COLLATE=
latin1_swedish_ci;
```

7.4 系统环境搭建

7.4.1 创建项目

创建一个名称为 ssm_crud 的 Maven 项目，可参照 5.1.2 节操作步骤完成项目创建。

7.4.2 添加项目依赖

在 ssm_crud 项目的 pom.xml 文件中添加 Spring 相关包、MyBatis 包、数据库驱动包、Servlet 包等依赖包，可参照 5.1.2 节操作步骤 1，添加依赖的 pom.xml 文件如下所示：

pom.xml

```
<project xmlns="http://maven.apache.org/POM/4.0.0"
xmlns:xsi="http://www.w3.org/2001/XMLSchema-instance"
    xsi:schemaLocation="http://maven.apache.org/POM/4.0.0
                        http://maven.apache.org/xsd/maven-4.0.0.xsd">
    <modelVersion>4.0.0</modelVersion>
    <groupId>com.crud</groupId>
    <artifactId>ssm_crud</artifactId>
    <version>0.0.1-SNAPSHOT</version>
    <packaging>war</packaging>
    <properties>
        <project.build.sourceEncoding>UTF-8</project.build.sourceEncoding>
        <maven.compiler.source>8</maven.compiler.source>
        <maven.compiler.target>8</maven.compiler.target>
    </properties>
    <dependencies>
        <dependency>
            <groupId>org.springframework</groupId>
```

```
        <artifactId>spring-webmvc</artifactId>
        <version>5.3.24</version>
    </dependency>
    <dependency>
        <groupId>org.springframework</groupId>
        <artifactId>spring-jdbc</artifactId>
        <version>5.3.24</version>
    </dependency>
    <dependency>
        <groupId>org.springframework</groupId>
        <artifactId>spring-aspects</artifactId>
        <version>5.3.24</version>
    </dependency>
    <dependency>
        <groupId>org.mybatis</groupId>
        <artifactId>mybatis</artifactId>
        <version>3.5.10</version>
    </dependency>
    <dependency>
        <groupId>org.mybatis</groupId>
        <artifactId>mybatis-spring</artifactId>
        <version>2.0.7</version>
    </dependency>
    <dependency>
        <groupId>c3p0</groupId>
        <artifactId>c3p0</artifactId>
        <version>0.9.1.2</version>
    </dependency>
    <dependency>
        <groupId>mysql</groupId>
        <artifactId>mysql-connector-java</artifactId>
        <version>5.1.49</version>
    </dependency>
    <dependency>
        <groupId>jstl</groupId>
        <artifactId>jstl</artifactId>
        <version>1.2</version>
    </dependency>
    <dependency>
        <groupId>javax.servlet</groupId>
```

```
            <artifactId>javax.servlet-api</artifactId>
            <version>3.0.1</version>
            <scope>provided</scope>
        </dependency>
        <dependency>
            <groupId>junit</groupId>
            <artifactId>junit</artifactId>
            <version>4.8.2</version>
        </dependency>
        <dependency>
            <groupId>org.springframework</groupId>
            <artifactId>spring-test</artifactId>
            <version>5.3.24</version>
            <scope>test</scope>
        </dependency>
        <dependency>
            <groupId>com.fasterxml.jackson.core</groupId>
            <artifactId>jackson-databind</artifactId>
            <version>2.9.2</version>
        </dependency>
        <dependency>
            <groupId>org.springframework</groupId>
            <artifactId>spring-aop</artifactId>
            <version>5.0.8.RELEASE</version>
        </dependency>
        <dependency>
            <groupId>javax</groupId>
            <artifactId>javaee-api</artifactId>
            <version>7.0</version>
            <scope>provided</scope>
        </dependency>
        <dependency>
            <groupId>com.alibaba</groupId>
            <artifactId>fastjson</artifactId>
            <version>1.2.41</version>
        </dependency>
    </dependencies>
    <build>
        <resources>
            <resource>
```

```
            <directory>src/main/resources</directory>
            <includes>
                <include>**/*.properties</include>
                <include>**/*.xml</include>
            </includes>
            <filtering>true</filtering>
        </resource>
        <resource>
            <directory>src/main/java</directory>
            <includes>
                <include>**/*.properties</include>
                <include>**/*.xml</include>
            </includes>
            <filtering>true</filtering>
        </resource>
    </resources>
  </build>
</project>
```

7.4.3　添加包

在 ssm_crud 项目中添加 pojo 包放置数据库实例类，添加 config 包放置连接配置类，添加 controller 包放置控制器处理类，添加 dao 包放置实体类，添加 service 包放置服务类，添加 util 包放置工具处理类，项目结构如图 7-1 所示。

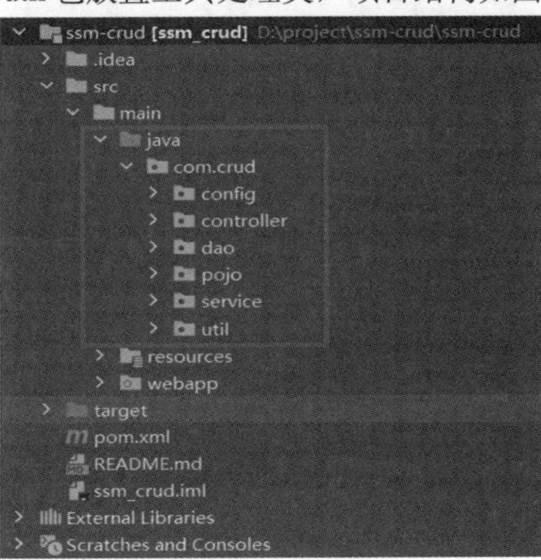

图 7-1　项目结构

图 7-1 中的 java 目录是源代码目录，包括一些类文件，开发人员根据项目实际需要创建目录，在每个目录中创建相关同类型的类。例如：将项目中工具类相关的代码放在 util 目录中；resources 目录是资源目录，包含项目的配置文件、属性文件等；target 目录是 Maven 构建的输出目录，通常包含编译后的类文件、打

包后的 jar 或 war 文件等。

7.5 模块开发

7.5.1 数据采集模块

1. 创建 POJO 相关类

在 ssm_crud 项目的 pojo 包中创建 Weather 类，类中 id 属性对应数据库主键 id， date 属性为日期，region 属性为地区，sunshine 属性为光照时长，avgwind 属性为平均风速，maxtemp 属性为最高温度，mintemp 属性为最低温度，avgtemp 属性为平均温度，envhumid 属性为湿度，avgpress 属性为平均气压，precipitation 属性为降雨量。结合属性的含义和使用场景，Weather 类(Weather.java)中规定了每个属性的访问范围和类型。

Weather.java

```
package com.crud.pojo;
public class Weather {
    private   int id ;
    private String date;
    private String region;
    private double sunshine;
    private double avgwind;
    private double maxtemp;
    private double mintemp;
    private double avgtemp;
    private double envhumid;
    private float avgpress;
    private double precipitation;
    public int getId() {
        return id;
    }
    public void setId(int id) {
        this.id = id;
    }
    public String getDate() {
        return date;
    }
    public void setDate(String date) {
        this.date = date;
    }
```

```
    public String getRegion() {
        return region;
    }
    public void setRegion(String region) {
        this.region = region;
    }
    public double getSunshine() {
        return sunshine;
    }
    public void setSunshine(double sunshine) {
        this.sunshine = sunshine;
    }
    public double getAvgwind() {
        return avgwind;
    }
    public void setAvgwind(double avgwind) {
        this.avgwind = avgwind;
    }
    public double getMaxtemp() {
        return maxtemp;
    }
    public void setMaxtemp(double maxtemp) {
        this.maxtemp = maxtemp;
    }
    public double getMintemp() {
        return mintemp;
    }
    public void setMintemp(double mintemp) {
        this.mintemp = mintemp;
    }
    public double getAvgtemp() {
        return avgtemp;
    }
    public void setAvgtemp(double avgtemp) {
        this.avgtemp = avgtemp;
    }
    public double getEnvhumid() {
        return envhumid;
    }
    public void setEnvhumid(double envhumid) {
```

```
        this.envhumid = envhumid;
    }
    public float getAvgpress() {
        return avgpress;
    }
    public void setAvgpress(float avgpress) {
        this.avgpress = avgpress;
    }
    public double getPrecipitation() {
        return precipitation;
    }
    public void setPrecipitation(double precipitation) {
        this.precipitation = precipitation;
    }
    @Override
    public String toString() {
        return "Weather{" +
                "id=" + id +
                ", date='" + date + '\'' +
                ", region='" + region + '\'' +
                ", sunshine=" + sunshine +
                ", avgwind=" + avgwind +
                ", maxtemp=" + maxtemp +
                ", mintemp=" + mintemp +
                ", avgtemp=" + avgtemp +
                ", envhumid=" + envhumid +
                ", avgpress=" + avgpress +
                ", precipitation=" + precipitation +
                '}';
    }
}
```

在 ssm_crud 项目的 POJO 包中创建 Yield 类(Yield.java)，其中 id 属性对应数据库主键 id，year 属性为年份，region 属性为地区，crop 属性为农作物名称，yield 属性为产量(单位为万吨)，area 属性为种植面积(单位为万亩)。

Yield.java

```
package com.crud.pojo;
public class Yield {
    private   int id ;
    private String year;
    private String region;
```

```
    private String crop;
    private double yield;
    private double area;
    public int getId() {
        return id;
    }
    public void setId(int id) {
        this.id = id;
    }
    public String getYear() {
        return year;
    }
    public void setYear(String year) {
        this.year = year;
    }
    public String getRegion() {
        return region;
    }
    public void setRegion(String region) {
        this.region = region;
    }
    public String getCrop() {
        return crop;
    }
    public void setCrop(String crop) {
        this.crop = crop;
    }
    public double getYield() {
        return yield;
    }
    public void setYield(double yield) {
        this.yield = yield;
    }
    public double getArea() {
        return area;
    }
    public void setArea(double area) {
        this.area = area;
}
    @Override
```

```
        public String toString() {
            return "Yield{" +
                    "id=" + id +
                    ", year='" + year + '\'' +
                    ", region='" + region + '\'' +
                    ", crop='" + crop + '\'' +
                    ", yield=" + yield +
                    ", area=" + area +
                    '}';
        }
    }
```

2. 创建 DAO 类

在 ssm_crud 项目的 dao 包中新建 WeatherDao 接口(WeatherDao.java)，接口中创建新增数据和查询数据的方法。

WeatherDao.java

```
package com.crud.dao;
import com.crud.pojo.Weather;
import java.util.List;
public interface WeatherDao {
        //新增数据
        void inertWeather(Weather weather);
       //获取全部数据
        List<Weather> selectAllWeather();
}
```

在 ssm_crud 项目的 dao 包中新建 YieldDao 接口(YieldDao.java)，接口中创建新增数据和查询数据的方法。

YieldDao.java

```
package com.crud.dao;
import com.crud.pojo.Yield;
import java.util.List;
public interface YieldDao {
    //获取全部数据
    List<Yield> selectAllYield();
    //新增数据
    void inertYield(Yield yield);
}
```

3. 创建 mapper 文件

在 ssm_crud 项目的 resources 目录下创建 mapper 包，用来放置映射文件，映射文件中字段名和类型与 MySQL 数据库中表字段名一致，数据库中表的字段名

可以参考 7.3 节。需要特别注意的是，在创建的 mapper 映射文件中添加 SQL 语句时，SQL 语句中的 id 字段值需要和步骤 2.创建 DAO 类中接口的方法名一致。在 mapper 包中新建 WeatherMapper.xml 文件、YieldMapper.xml 文件。

WeatherMapper.xml

```
<?xml version="1.0" encoding="UTF-8"?>
<!DOCTYPE mapper PUBLIC "-//mybatis.org//DTD Mapper 3.0//EN"
"http://mybatis.org/dtd/mybatis-3-mapper.dtd">
<mapper namespace="com.crud.dao.WeatherMapper">
    <resultMap id="WeatherResultMap" type="com.crud.pojo.Weather">
        <id column="id" jdbcType="INTEGER" property="id" />
        <result column="date" jdbcType="VARCHAR" property="date" />
        <result column="region" jdbcType="VARCHAR" property="region" />
        <result column="sunshine" jdbcType="FLOAT" property="sunshine" />
        <result column="avgwind" jdbcType="FLOAT" property="avgwind" />
        <result column="maxtemp" jdbcType="FLOAT" property="maxtemp" />
        <result column="mintemp" jdbcType="FLOAT" property="mintemp" />
        <result column="avgtemp" jdbcType="FLOAT" property="avgtemp" />
        <result column="envhumid" jdbcType="FLOAT" property="envhumid" />
        <result column="avgpress" jdbcType="FLOAT" property="avgpress" />
        <result column="precipitation" jdbcType="FLOAT" property="precipitation" />
    </resultMap>
    <select id="selectAllWeather" resultMap="WeatherResultMap">
        select * from weather
    </select>
    <insert id="inertWeather" parameterType="com.crud.pojo.Weather">
        insert into weather (id, date,  region,  sunshine, avgwind,  maxtemp, mintemp,
        avgtemp, envhumid, avgpress, precipitation)
        values (#{id,jdbcType=INTEGER}, #{date,jdbcType=VARCHAR},
                #{region,jdbcType=VARCHAR}, #{sunshine, jdbcType=FLOAT},
                #{avgwind, jdbcType=FLOAT}, #{maxtemp, jdbcType=FLOAT},
                #{mintemp, jdbcType=FLOAT}, #{avgtemp, jdbcType=FLOAT},
                #{envhumid, jdbcType=FLOAT}, #{avgpress, jdbcType=FLOAT},
                #{precipitation, jdbcType=FLOAT})
    </insert>
</mapper>
```

上述文件中定义了数据库中 Weather 表的字段名称和字段类型、Weather 表映射到 Weather 类的属性名称、一个查询语句和一个插入语句。查询语句和插入语句的 id 字段值分别与 WeatherDao 接口中的两个方法的方法名一致。

YieldMapper.xml

```
<?xml version="1.0" encoding="UTF-8"?>
```

```
<!DOCTYPE mapper PUBLIC "-//mybatis.org//DTD Mapper 3.0//EN" "http://mybatis.org/dtd/
mybatis-3-mapper.dtd">
<mapper namespace="com.crud.dao.YieldMapper">
    <resultMap id="YieldResultMap" type="com.crud.pojo.Yield">
        <id column="id" jdbcType="INTEGER" property="id" />
        <result column="year" jdbcType="VARCHAR" property="year" />
        <result column="region" jdbcType="VARCHAR" property="region" />
        <result column="crop" jdbcType="VARCHAR" property="crop" />
        <result column="yield" jdbcType="FLOAT" property="yield" />
        <result column="area" jdbcType="FLOAT" property="area" />
    </resultMap>
    <select id="selectAllYield" resultMap="YieldResultMap">
        select * from yield
    </select>
    <insert id="inertYield" parameterType="com.crud.pojo.Yield">
        insert into yield (id, year, region, crop,yield, area)
        values (#{id, jdbcType=INTEGER}, #{year, jdbcType=VARCHAR},
                #{region, jdbcType=VARCHAR}, #{crop, jdbcType=VARCHAR},
                #{yield, jdbcType=FLOAT}, #{area, jdbcType=FLOAT})
    </insert>
</mapper>
```

上述文件中定义了数据库中 Yield 表的字段名称和字段类型、Yield 表映射到 Yield 类的属性名称、一个查询语句和一个插入语句。查询语句和插入语句的 id 字段值分别与 YieldDao 接口中的两个方法的方法名一致。

7.5.2 数据存储模块

数据库中的原始数据无法直接使用，需要对数据进行处理后才能用于前端页面的展示。在 ssm_crud 项目的 pojo 目录下新增 ShowWeather 类、WindData 类、PrecipData 类、UiBean 类；在 ssm_crud 项目的 service 目录下新增 WeatherService 类。

1. 添加数据存储相关类

ShowWeather 类主要用于返回展示数据给前端，ShowWeather.java 文件中的 avgSunShine 属性为月均日照时长，avgAvgwind 属性为月均风速，avgMaxtemp 属性为月均最高温度，avgMintemp 属性为月均最低温度，avgPrecipitation 属性为月均降雨量，文件中还定义了 getter 方法和 setter 方法。

ShowWeather.java

```
package com.crud.pojo;
public class ShowWeather {
    private double avgSunShine;
    private double avgAvgwind;
```

```
    private double avgMaxtemp;
    private double avgMintemp;
    private double avgPrecipitation;
    public double getAvgSunShine() {
        return avgSunShine;
    }
    public void setAvgSunShine(double avgSunShine) {
        this.avgSunShine = avgSunShine;
    }
    public double getAvgAvgwind() {
        return avgAvgwind;
    }
    public void setAvgAvgwind(double avgAvgwind) {
        this.avgAvgwind = avgAvgwind;
    }
    public double getAvgMaxtemp() {
        return avgMaxtemp;
    }
    public void setAvgMaxtemp(double avgMaxtemp) {
        this.avgMaxtemp = avgMaxtemp;
    }
    public double getAvgMintemp() {
        return avgMintemp;
    }
    public void setAvgMintemp(double avgMintemp) {
        this.avgMintemp = avgMintemp;
    }
    public double getAvgPrecipitation() {
        return avgPrecipitation;
    }
    public void setAvgPrecipitation(double avgPrecipitation) {
        this.avgPrecipitation = avgPrecipitation;
    }
}
```

WindData 类(WindData.java)用于封装返回的风速数据，其中 name 属性为月份名称，value 属性为月均风速值，类中还定义了每个属性的 getter 和 setter 方法。

WindData.java

```
package com.crud.pojo;
public class WindData {
    private String name;
```

```
    private double value;
    public String getName() {
        return name;
    }
    public void setName(String name) {
        this.name = name;
    }
    public double getValue() {
        return value;
    }
    public void setValue(double value) {
        this.value = value;
    }
    @Override
    public String toString() {
        return "WindData{" +
                "name='" + name + '\'' +
                ", value=" + value +
                '}';
    }
}
```

PrecipData 类(PrecipData.java)用于封装返回的风速数据，其中 name 属性为月份名称，value 属性为月均降雨量值。类中还定义了每个属性的 getter 和 setter 方法。

PrecipData.java

```
package com.crud.pojo;
public class PrecipData {
    private String name;
    private double value;
    public String getName() {
        return name;
    }
    public void setName(String name) {
        this.name = name;
    }
    public double getValue() {
        return value;
    }
    public void setValue(double value) {
        this.value = value;
```

```
    }
    @Override
    public String toString() {
        return "PrecipData{" +
                "name='" + name + '\'' +
                ", value=" + value +
                '}';
    }
}
```

UiBean 类(UiBean.java)用于封装返回给前端的所有数据，其中 dateMonth 属性表示月份集合，sunshine 属性表示月均日照时长集合，avgwind 属性表示月均风速集合，avgMaxtemp 属性表示月均最高温度集合，avgMintemp 属性表示最低温度集合，avgPrecipitation 属性表示月均降雨量集合，regionName 属性表示地区名集合，yieldNum 属性表示种植农作物面积，areaNum 属性表示农作物产量集合。

UiBean.java

```
package com.crud.util;
import com.crud.pojo.PrecipData;
import com.crud.pojo.WindData;
import java.util.List;
public class UiBean {
    private String[] dateMonth;
    private double[] sunshine;
    private double[] avgwind;
    private List<WindData> windDataList;
    private List<PrecipData> preDataList;
    private double[] avgMaxtemp;
    private double[] avgMintemp;
    private double[] avgPrecipitation;
    private String[] regionName;
    private double[] yieldNum;
    private double[] areaNum;
    public  UiBean(){
    }
    public String[] getDateMonth() {
        return dateMonth;
    }
    public void setDateMonth(String[] dateMonth) {
        this.dateMonth = dateMonth;
    }
    public double[] getSunshine() {
```

```
        return sunshine;
    }
    public void setSunshine(double[] sunshine) {
        this.sunshine = sunshine;
    }
    public double[] getAvgwind() {
        return avgwind;
    }
    public void setAvgwind(double[] avgwind) {
        this.avgwind = avgwind;
    }
    public double[] getAvgMaxtemp() {
        return avgMaxtemp;
    }
    public void setAvgMaxtemp(double[] avgMaxtemp) {
        this.avgMaxtemp = avgMaxtemp;
    }
    public double[] getAvgMintemp() {
        return avgMintemp;
    }
    public void setAvgMintemp(double[] avgMintemp) {
        this.avgMintemp = avgMintemp;
    }
    public double[] getAvgPrecipitation() {
        return avgPrecipitation;
    }
    public void setAvgPrecipitation(double[] avgPrecipitation) {
        this.avgPrecipitation = avgPrecipitation;
    }
    public List<WindData> getWindDataList() {
        return windDataList;
    }
    public void setWindDataList(List<WindData> windDataList) {
        this.windDataList = windDataList;
    }
    public List<PrecipData> getPreDataList() {
        return preDataList;
    }
    public void setPreDataList(List<PrecipData> preDataList) {
        this.preDataList = preDataList;
```

```
    }
    public String[] getRegionName() {
        return regionName;
    }
    public void setRegionName(String[] regionName) {
        this.regionName = regionName;
    }
    public double[] getYieldNum() {
        return yieldNum;
    }
    public void setYieldNum(double[] yieldNum) {
        this.yieldNum = yieldNum;
    }
    public double[] getAreaNum() {
        return areaNum;
    }
    public void setAreaNum(double[] areaNum) {
        this.areaNum = areaNum;
    }
}
```

2. 添加数据处理相关类

WeatherService 类实现数据处理逻辑，WeatherService.java 文件中定义了数据处理的方法，其中 inertWeather()方法用于生成模拟数据并将数据存入数据库中，getWeather()方法获取温度、湿度、光照时长、风速、降雨量等数据，getYield()方法获取农作物产区、种植面积、产量等数据，getAll()方法将获取的所有数据进行组合并转化为 JSON 格式的数据。

WeatherService.java

```
package com.crud.service;
import com.crud.pojo.*;
import com.crud.dao.WeatherMapper;
import com.crud.dao.YieldMapper;
import org.springframework.beans.factory.annotation.Autowired;
import org.springframework.stereotype.Service;
import org.springframework.transaction.annotation.Transactional;
import java.math.BigDecimal;
import java.text.SimpleDateFormat;
import java.util.*;
@Service("weatherService")
@Transactional
public class WeatherService {
```

```
@Autowired
WeatherMapper weatherMapper;
@Autowired
YieldMapper yieldMapper;
//插入数据的方法
public void  inertWeather(){
    List<Weather> weatherList = new ArrayList<Weather>();
    //生成 2023 年的日期，从 2023 年 1 月 1 日至 2023 年 12 月 31 日
    Calendar calendar = Calendar.getInstance();
    calendar.set(Calendar.YEAR, 2023);
    SimpleDateFormat sdf = new SimpleDateFormat("yyyy-MM-dd");
    int totalDays = calendar.getActualMaximum(Calendar.DAY_OF_YEAR);
    for (int i = 1; i <= totalDays; i++) {
        Weather weather = new Weather();
        weather.setRegion("某市");
        calendar.set(Calendar.DAY_OF_YEAR, i);
        Date date = calendar.getTime();
        String dateFormatStr = sdf.format(date);
        weather.setDate(dateFormatStr);
        Random random = new Random();
        //光照时长数
        int min = 6;
        int max = 10;
        int randomSun = random.nextInt((max - min + 1)) + min;
        weather.setSunshine(randomSun);
        //平均风速(m/s)
        int mins = 6;
        int maxs = 10;
        int randomWind = random.nextInt((maxs - mins + 1)) + mins;
        weather.setAvgwind(randomWind);
        //最高温度
        double mint = 15.5;
        double maxt = 42.1;
        double randomTemp = mint + (maxt - mint) * random.nextDouble();
        BigDecimal bd = new BigDecimal(randomTemp);
        double randomTempa = bd.setScale(1, BigDecimal.ROUND_DOWN).doubleValue();
        weather.setMaxtemp(randomTempa);
        //最低温度
        double mine = -10;
        double maxe =  13;
```

```
            double randomTemi= mine + (maxe - mine) * random.nextDouble();
            BigDecimal ba = new BigDecimal(randomTemi);
            double randomTemia = ba.setScale(1, BigDecimal.ROUND_DOWN).doubleValue();
            weather.setMintemp(randomTemia);
            //平均温度
            double avgTemp = (randomTempa + randomTemia)/2;
            BigDecimal be = new BigDecimal(avgTemp);
            double randomTeme = be.setScale(1, BigDecimal.ROUND_DOWN).doubleValue();
            weather.setAvgtemp(randomTeme);
            //环境湿度
            double minh = 50;
            double maxh = 80;
            double randomHuan= minh + (maxh - minh) * random.nextDouble();
            BigDecimal bf = new BigDecimal(randomHuan);
            double randomTems = bf.setScale(1, BigDecimal.ROUND_DOWN).doubleValue();
            weather.setEnvhumid(randomTems);
            //平均气压
            int minp = 950;
            int maxp = 1050;
            int randomPress = random.nextInt((maxp - minp + 1)) + minp;
            weather.setAvgpress(randomPress);
            //降雨量
            int minj = 50;
            int maxj = 150;
            int randomPrecip = random.nextInt((maxj - minj + 1)) + minj;
            weather.setPrecipitation(randomPrecip);
            weatherList.add(weather);
        }
        //数据入库
        for (Weather w: weatherList) {
            weatherMapper.inertWeather(w);
        }
    }
}
```

7.5.3　数据处理模块

1. 添加控制器类

在 ssm_crud 项目的 controller 包中创建 WeatherController 类(WeatherController.java)，定义了两个方法响应视图层的请求。控制器收到请求后再调用服务层相关

的接口完成对数据的操作。例如，当视图层接收到查询数据的请求时，会执行 getAllWeather()方法，调用 WeatherService 类中查询数据的方法，该控制器类为 WeatherController. java。

WeatherController.java

```
package com.crud.controller;
import com.crud.service.WeatherService;
import org.springframework.beans.factory.annotation.Autowired;
import org.springframework.stereotype.Controller;
import org.springframework.web.bind.annotation.*;
@Controller
@CrossOrigin
public class WeatherController {
    @Autowired
    WeatherService weatherService;
    // 插入数据的方法
    @RequestMapping("/insertData")
    @ResponseBody
    public void getWithJson() {
        weatherService.inertWeather();
    }
    //查询所有数据的方法
    @RequestMapping("/getAllData")
    @ResponseBody
    public void getAllWeather(){
        weatherService.getAll();
    }
}
```

2. 添加通信类

在 ssm_crud 项目的 config 包中创建 UiWebSocket 通信类(UiWebSocket.java)，该类实现服务连接、消息发送、消息接收、连接验证等功能。

UiWebSocket.java

```
package com.crud.config;
import com.crud.service.WeatherService;
import org.springframework.context.ApplicationContext;
import org.springframework.context.support.ClassPathXmlApplicationContext;
import javax.websocket.OnClose;
import javax.websocket.OnError;
import javax.websocket.OnOpen;
import javax.websocket.Session;
```

```
import javax.websocket.server.ServerEndpoint;
import java.io.IOException;
import java.util.concurrent.CopyOnWriteArraySet;
//将当前类定义成一个 WebSocket 服务器
@ServerEndpoint("/uiwebSocket")
public class UiWebSocket {
    // 连接数
    private static  int onlineCount=0;
    private static CopyOnWriteArraySet<UiWebSocket> webSockets=
            new CopyOnWriteArraySet<UiWebSocket>();
    // 连会话数
    private Session session;
    // 发送消息
    public  void sendMessage(String msg){
        try {
            this.session.getBasicRemote().sendText(msg);
        } catch (IOException e) {
            e.printStackTrace();
        }
    }
    // 获取连接数
    public static synchronized int getOnlineCount(){
        return onlineCount;
    }
    //增加连接数
    public static synchronized void   addOnlineCount(){
        UiWebSocket.onlineCount++;
    }
    public static synchronized void subOnlineCount(){
        UiWebSocket.onlineCount--;
    }
    //使用 weatherService 服务
    WeatherService weatherService;
    //收到消息
    public void onMessage(String msg, Session session){
        System.out.println("来自客户端信息:"+msg);
        for(final UiWebSocket uiWebSocket:webSockets){
            while (true){
                //从数据库获取数据发送到消息队列
                uiWebSocket.sendMessage(weatherService.getAll());
```

```
            }
        }
    }
    //连接成功
    @OnOpen
    public void onOpen(Session session){
        this.session=session;
        webSockets.add(this);
        addOnlineCount();
        System.out.println("有新的连接，当前连接数为"+getOnlineCount());
        ApplicationContext context=
                    new ClassPathXmlApplicationContext("applicationContext.xml");
        weatherService = context.getBean("weatherService", WeatherService.class);
        onMessage("", session);
    }
    //关闭
    @OnClose
    public void onClose( ){
        webSockets.remove(this);
        subOnlineCount();
        System.out.println("关闭一个连接后在线人数："+getOnlineCount());
    }
    //出错
    @OnError
    public  void onError(Session session, Throwable throwable){
        System.out.println("发生错误");
        throwable.printStackTrace();
    }
}
```

上述代码中，onOpen()方法是与客户端建立连接时的回调方法，session 用于识别连接信息，sendMessage()方法用于发送消息，onMessage()方法是接收到来客户端请求时的回调方法，onClose()方法在 WebSocket 连接关闭时被触发，onError()方法在 WebSocket 连接发生错误时被触发。

3. 创建 spring 相关配置文件

在 ssm_crud 项目的 resources 目录下创建数据库配置文件 db.properties，用于配置数据源，包括数据库驱动、数据库地址、用户名、密码等。

db.properties

```
jdbc.driver=com.mysql.jdbc.Driver
jdbc.url=jdbc:mysql://localhost:3306/mysql?characterEncoding=utf8
jdbc.username=root
```

```
jdbc.password=123
jdbc.maxActive=30
jdbc.initialSize=20
jdbc.maxWait=500
```

在 ssm_crud 项目的 resources 目录下创建 MyBatis 的核心配置文件(mybatis-config.xml)，配置命名规则、实体类位置等。

mybatis-config.xml

```
<?xml version="1.0" encoding="UTF-8" ?>
<!DOCTYPE configuration
PUBLIC "-//mybatis.org//DTD Config 3.0//EN"
"http://mybatis.org/dtd/mybatis-3-config.dtd">
<configuration>
    <!-- 开启驼峰命名规则 -->
    <settings>
        <setting name="mapUnderscoreToCamelCase" value="true"/>
    </settings>
    <!-- 配置别名，实体类的位置，方便引用-->
    <typeAliases>
        <package name="com.crud.dao"/>
    </typeAliases>
    <plugins>
        <plugin interceptor="com.github.pagehelper.PageInterceptor">
            <property name="reasonable" value="true"/>
            </plugin>
    </plugins>
</configuration>
```

在 ssm_crud 项目的 resources 目录下创建 springMVC 配置文件(springmvc.xml)，配置要扫描的控制器路径、视图解析器等。

springmvc.xml

```
<?xml version="1.0" encoding="UTF-8"?>
<beans xmlns="http://www.springframework.org/schema/beans"
    xmlns:xsi="http://www.w3.org/2001/XMLSchema-instance"
    xmlns:p="http://www.springframework.org/schema/p"
    xmlns:context="http://www.springframework.org/schema/context"
    xmlns:mvc="http://www.springframework.org/schema/mvc"
    xsi:schemaLocation="http://www.springframework.org/schema/beans
    http://www.springframework.org/schema/beans/spring-beans-4.0.xsd
        http://www.springframework.org/schema/mvc
        http://www.springframework.org/schema/mvc/spring-mvc-4.0.xsd
        http://www.springframework.org/schema/context
```

```
        http://www.springframework.org/schema/context/spring-context-4.0.xsd">
    <!-- 1. 扫描业务逻辑组件，配置组件扫描器，注解式使用，只扫描控制器-->
    <context:component-scan base-package="com.crud.controller" />
    <!-- 2. 配置视图解析器-->
    <bean class="org.springframework.web.servlet.view.InternalResourceViewResolver">
        <!-- 前缀 -->
        <property name="prefix" value="/WEB-INF/views/" />
        <!-- 后缀 -->
        <property name="suffix" value=".jsp" />
    </bean>
<mvc:default-servlet-handler/>
    <!-- 3. 配置注解驱动，映射动态请求，支持 springmvc 一些高级功能，比如 JSR303
校验，快捷 AJAX 请求-->
    <mvc:annotation-driven/>
</beans>
```

在 resources 目录下创建配置文件(application.xml)，配置数据源、扫描器、aop 等。

application.xml

```
<?xml version="1.0" encoding="UTF-8"?>
<?xml version="1.0" encoding="UTF-8"?>
<beans xmlns:xsi="http://www.w3.org/2001/XMLSchema-instance"
    xmlns="http://www.springframework.org/schema/beans"
    xmlns:context="http://www.springframework.org/schema/context"
    xmlns:aop="http://www.springframework.org/schema/aop"
    xmlns:tx="http://www.springframework.org/schema/tx"
    xsi:schemaLocation="http://www.springframework.org/schema/beans
    http://www.springframework.org/schema/beans/spring-beans-4.2.xsd
    http://www.springframework.org/schema/context
    http://www.springframework.org/schema/context/spring-context-4.2.xsd
    http://www.springframework.org/schema/aop
    http://www.springframework.org/schema/aop/spring-aop-4.2.xsd
    http://www.springframework.org/schema/tx
    http://www.springframework.org/schema/tx/spring-tx.xsd">
    <!-- 1. 配置读取 jdbc.properties 文件-->
    <context:property-placeholder location="classpath:jdbc.properties"/>
    <!-- 2. 配置 C3P0 连接池 -->
    <bean name="C3P0DataSource" class="com.mchange.v2.c3p0.ComboPooledDataSource" >
        <property name="jdbcUrl" value="${jdbc.url}" ></property>
        <property name="driverClass" value="${jdbc.driver}" ></property>
        <property name="user" value="${jdbc.username}" ></property>
        <property name="password" value="${jdbc.password}" ></property>
    </bean>
```

```
    <!-- Service 的注册业务逻辑注册-->
    <!-- 3. 配置扫描器-->
    <context:component-scan base-package="com.crud.service" />
    <!-- 4. 注册 SqlSessionFactoryBean -->
    <bean id="sqlSessionFactory" class="org.mybatis.spring.SqlSessionFactoryBean">
        <!-- 4.1 指定 mybatis 主配置文件的位置-->
        <property name="configLocation" value="classpath:mybatis.xml"/>
        <!-- 4.2 连接池注入 -->
        <property name="dataSource" ref="C3P0DataSource"/>
        <!--4.3 指定 mapper 文件的位置-->
        <property name="mapperLocations" value="classpath:mapper/*.xml"/>
    </bean>
    <!-- 5. 配置扫描器 -->
    <bean class="org.mybatis.spring.mapper.MapperScannerConfigurer">
        <property name="sqlSessionFactoryBeanName" value="sqlSessionFactory"/>
        <!-- 5.1 扫描所有的 DAO 接口并加入到 IOC 容器中 -->
        <property name="basePackage" value="com.crud.dao" />
    </bean>
    <bean id="sqlSession" class="org.mybatis.spring.SqlSessionTemplate">
        <constructor-arg name="sqlSessionFactory" ref="sqlSessionFactory" />
        <constructor-arg name="executorType" value="BATCH" />
    </bean>
    <!-- 6. 配置事务管理器 -->
    <bean id="transactionManager"
        class="org.springframework.jdbc.datasource.DataSourceTransactionManager">
        <property name="dataSource" ref="C3P0DataSource"/>
    </bean>
    <!-- 7. 注册事务通知 -->
    <tx:advice id="transactionAdvice" transaction-manager="transactionManager">
        <tx:attributes>
            <tx:method name="*" isolation="DEFAULT" propagation="REQUIRED"/>
            <tx:method name="get*" read-only="true"/>
        </tx:attributes>
    </tx:advice>
    <!-- 8. 配置 aop -->
    <aop:config>
        <aop:pointcut expression="execution(* com.crud.service..*(..))" id="myPointCut"/>
        <aop:advisor advice-ref="transactionAdvice" pointcut-ref="myPointCut"/>
    </aop:config>
</beans>
```

4. 添加 web.xml

在 ssm_crud 项目的 WEB-INF 文件夹下创建配置文件 web.xml，可参照 5.1.2 节

中的步骤 4。

web.xml

```
<?xml version="1.0" encoding="UTF-8"?>
<web-app xmlns:xsi="http://www.w3.org/2001/XMLSchema-instance"
    xmlns="http://java.sun.com/xml/ns/javaee"
    xsi:schemaLocation="http://java.sun.com/xml/ns/javaee
                    http://java.sun.com/xml/ns/javaee/web-app_2_5.xsd"
    id="WebApp_ID" version="2.5">
    <!-- 1.注册 Spring 配置文件的位置-->
    <context-param>
        <param-name>contextConfigLocation</param-name>
        <param-value>classpath:applicationContext.xml</param-value>
    </context-param>
    <!-- 2.注册 ServletContext 监听器-->
    <listener>
    <listener-class>org.springframework.web.context.ContextLoaderListener</listener-class>
    </listener>
    <!-- 3. 注册字符集过滤器一定要放在所有过滤器的最前面-->
    <filter>
        <filter-name>CharacterEncodingFilter</filter-name>
    <filter-class>org.springframework.web.filter.CharacterEncodingFilter</filter-class>
        <init-param>
            <param-name>encoding</param-name>
            <param-value>utf-8</param-value>
        </init-param>
        <init-param>
            <param-name>forceRequestEncoding</param-name>
            <param-value>true</param-value>
        </init-param>
        <init-param>
            <param-name>forceReponseEncoding</param-name>
            <param-value>true</param-value>
        </init-param>
    </filter>
    <filter-mapping>
        <filter-name>CharacterEncodingFilter</filter-name>
        <url-pattern>/*</url-pattern>
    </filter-mapping>
    <!-- 4. 配置中央调度器用于拦截所有请求-->
```

```
    <servlet>
        <servlet-name>springmvc</servlet-name>
        <servlet-class>org.springframework.web.servlet.DispatcherServlet</servlet-class>
        <init-param>
            <param-name>contextConfigLocation</param-name>
            <param-value>classpath:spring-mvc.xml</param-value>
        </init-param>
        <load-on-startup>1</load-on-startup>
    </servlet>
    <servlet-mapping>
        <servlet-name>springmvc</servlet-name>
        <url-pattern>/</url-pattern>
    </servlet-mapping>
    <filter>
        <filter-name>HiddenHttpMethodFilter</filter-name>
    <filter-class>org.springframework.web.filter.HiddenHttpMethodFilter</filter-class>
    </filter>
    <filter-mapping>
        <filter-name>HiddenHttpMethodFilter</filter-name>
        <url-pattern>/*</url-pattern>
    </filter-mapping>
    <filter>
        <filter-name>HttpPutFormContentFilter</filter-name>
    <filter-class>org.springframework.web.filter.HttpPutFormContentFilter</filter-class>
    </filter>
    <filter-mapping>
        <filter-name>HttpPutFormContentFilter</filter-name>
        <url-pattern>/*</url-pattern>
    </filter-mapping>
</web-app>
```

7.5.4　数据可视化模块

1. 添加前端资源

在 ssm_crud 项目的 webapp 目录下创建 static 目录，该目录用于存放静态资源及 js 文件。在 static 目录下创建 bootstrap 和 js 目录，分别导入项目需要的 JAR 包，包括 echarts.js 和 jquery.js，导入后的效果图如图 7-2 所示。

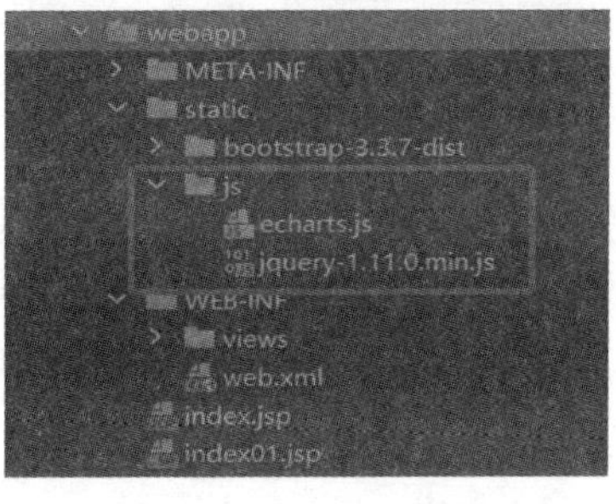

图 7-2　项目 JAR 包

2. 添加 jsp 文件

在 ssm_crud 项目的 webapp 目录下创建视图文件 index.jsp，并在文件中引入 ECharts 组件绘制柱状图、折线图、饼状图等反映农业生产状况的图表，从而实现农业数据的可视化。

index.jsp

```
<%@ page language="java" contentType="text/html; charset=UTF-8" pageEncoding="UTF-8"%>
<!DOCTYPE html PUBLIC "-//W3C//DTD HTML 4.01 Transitional//EN" "http://www.w3.org/
TR/html4/loose.dtd">
<html>
<head>
    <meta http-equiv="Content-Type" content="text/html; charset=UTF-8">
    <title>农业数据分析平台</title>
    <!-- 导入 JQuery -->
    <script type="text/javascript" src="static/js/jquery-1.11.0.min.js"></script>
    <!-- 导入 ECharts -->
    <script type="text/javascript" src="static/js/echarts.js" ></script>
    <% pageContext.setAttribute("APP_PATH", request.getContextPath()); %>
</head>
<body>
<div style="height:30px ">
    <div style="float:left;height:30px;font-style: normal;font-size: 30px;font-weight: 500;
    margin-bottom: 20px;/* color: blue; */">农业数据分析平台</div>
    <div style="height: 30px;float: right" id="timeDiv"></div>
</div>
<div style="display: flex;justify-content: center">
    <div>
        <div style="width: 1800px;height:1800px ">
            <div style="height: 1200px;width: 1800px">
                        <div style="width: 1000px;float: left;height: 1200px">
                    <%--1. 这个是用来展示日照数据的 div，这是一个柱状图--%>
                            <div id="sunshineDiv" style="height: 600px;width: 1000px;"></div>
                      <%--2. 这是一个展示高温和低温均值的折线图--%>
                      <div id="tempDiv" style="height: 500px;width: 1000px;"></div>
                </div>
                  <div style="width: 800px;height: 1200px;float: right">
                      <%--3. 这个是用来展示风速数据的 div，这是一个饼状图--%>
                      <div id="windDiv" style="height: 600px;width: 800px;"></div>
                      <%--4. 这是一个展示降雨量的环形图--%>
```

```
                    <div id="precipDiv" style="height: 600px;width: 800px;"></div>
                </div>
            </div>
            <%--5. 这是一个展示产量的柱状图--%>
            <div id="yieldDiv" style="height: 600px;width: 1800px;"></div>
        </div>
    </div>
</div>
<script type="text/javascript">
    $(function() {
        //渲染数据的方法
        echartShowa();
        //更新时间的方法
        updateTime();
    })
    setInterval(updateTime, 1000)
    //实时获取当前时间，并将数据填充到 id 为 timeDiv 的 div 标签中
    function updateTime() {
        var mydate = (new Date()).toLocaleString();
        var timeDiv = document.getElementById("timeDiv");
        timeDiv.textContent = mydate
        console.log(mydate)
    }
    //渲染日照时长、降雨量、风速、温度等数据
    function echartShowa(){
        // 1. 用来渲染日照数据
        var mychart=echarts.init(document.getElementById("sunshineDiv"));
        mychart.setOption({
            xAxis: {
                type: "category",
                name:"时间",
                data:[]
            },
            yAxis:{
                type: "value",
            },
            title:{
                text: "2023 年日照时长分析",
```

```
            subtext: "Demo 虚构数据",
            x: "center"
        },
        tooltip: {
            trigger: "axis",
            axisPointer: {
                type: "shadow"
            }
        },
        legend:{
            show:true
        },
        series:[
            {
                type: "bar",
                data:[],
                label: {
                    show: true,
                    position: "top"
                }
            }
        ]
    });
    mychart.hideLoading();
    // 2. 用来渲染风速饼状图
    var mychartWind=echarts.init(document.getElementById("windDiv"));
    mychartWind.setOption({
        title : {
            text : "2023 年风速分析",
            subtext: "Demo 虚构数据",
            x: "center"
        },
        legend:{
            show:false
        },
        series : [
            {
                name: "月均风速",
                type: "pie",
```

```
                label: {
                    show: true,
                    position: "outside",
                    formatter: "{b}({d}%)",
                },
                radius: "80%"',
                data:[]
            }
        ]
});
// 3. 用来渲染降雨量圆环图
var mychartPrecip=echarts.init(document.getElementById("precipDiv"));
mychartPrecip.setOption({
    title : {
        text : "2023 年降雨量分析",
        subtext: "Demo 虚构数据",
        x: "center"
    },
    legend:{
        show:false
    },
    series : [
        {
            name: "月均降雨量",
            type: "pie",
            label: {
                show: true,
                position: "outside",
                formatter: "{b}({d}%)",
            },
            radius:[ "60%", "80%"],
            data:[]
        }
    ]
});
// 4. 用来渲染高温和低温均值数据
var mychartTemp=echarts.init(document.getElementById("tempDiv"));
mychartTemp.setOption({
    title:{
```

```
        text: "2023 年温度分析",
        subtext: "Demo 虚构数据",
        x: "center"
    },
    tooltip: {
        trigger: "axis",
        axisPointer: {
            type: "shadow"
        }
    },
    legend:{
        orient: "horizontal",
        x: "left",
        y: "top",
        data: ["高温均值", "低温均值"]
    },
    grid: {
        top: "20%",
        left: "3%",
        right: "10%",
        bottom: "5%",
        containLabel: true
    },
    color: ["#3366CC", "#FFCC99"],
    xAxis: {
        type: "category",
        name:"时间",
        data:[]
    },
    yAxis:{
        type: "value",
    },
    series:[
        {
            type: "line",
            data:[],
            name: "高温均值",
            label: {
                show: true,
```

```
                position: "top"
            }
        },{
            type: "line",
            data:[],
            name: "低温均值",
            label: {
                show: true,
                position: "top"
            }
        }
    ]
});
// 5. 用来渲染各地区的产量
var mychartYield = echarts.init(document.getElementById("yieldDiv"));
mychartYield.setOption({
    xAxis: {
        type: "category",
        name:"地区",
        data:[]
    },
    yAxis:{
        type: "value",
    },
    title:{
        text: "2023 年某市各地区小麦产量",
        subtext: "Demo 虚构数据",
        x: "center"
    },
    tooltip: {
        trigger: "axis",
        axisPointer: {
            type: "shadow"
        }
    },
    legend:{
        orient: "horizontal",
        x: "left",
        y: "top",
```

```
            data: ["产量", "种植面积"]
        },
        series:[
            {
                type: "bar",
                name: "产量(万吨) ",
                data:[],
                label: {
                    show: true,
                    position: "top"
                }
            },{
                type: "bar",
                data:[],
                name: "种植面积(万亩) ",
                label: {
                    show: true,
                    position: "top"
                }
            }
        ]
    });
    var websocket=null;
    function closeWebsocket(){
        websocket.close();
    }
    if("WebSocket" in window){
            websocket=new WebSocket("ws://localhost:8080/ssm_crud_war/uiwebSocket");
    }
    else{
        alert("当前浏览器不支持 websocket");
    }
    websocket.onerror=function (){
        setMessageInnerHTML("websocket 连接发生错误");
    }
    websocket.onopen=function (){
        setMessageInnerHTML("websocket 连接成功");
    }
    websocket.onclose=function (){
```

```
        setMessageInnerHTML("websocket 断开连接");
    }
    websocket.onmessage=function (event) {
        jsonbean = JSON.parse(event.data);
        mydate = (new Date()).toLocaleString();
        // 1) 填充日照数据
        mychart.setOption({
            xAxis: {
                data: jsonbean.dateMonth
            },
            series: [{
                data: jsonbean.sunshine,
            }]
        });
        // 2) 填充风速数据
        mychartWind.setOption({
            legend:{
                data:jsonbean.dateMonth
            },
            series: [{
                data:jsonbean.windDataList,
            }]
        })
        // 3) 填充降雨量数据
        mychartPrecip.setOption({
            legend:{
                data:jsonbean.dateMonth
            },
            series: [{
                data:jsonbean.preDataList,
            }]
        })
        // 4) 填充温度数据
        mychartTemp.setOption({
            legend:{
                data:[ "高温均值", "低温均值"]
            },
            xAxis: {
                data: jsonbean.dateMonth
```

```
                },
                series: [{
                    data: jsonbean.avgMaxtemp
                },{
                    data:jsonbean.avgMintemp
                }]
            });
            // 5) 填充产量数据
            mychartYield.setOption({
                legend:{
                    data:[ "产量(万吨)", "种植面积(万亩)"]
                },
                xAxis: {
                    data: jsonbean.regionName
                },
                series: [{
                    data: jsonbean.yieldNum,
                },{
                    data: jsonbean.areaNum,
                }]
            });
            window.onbeforeunload = function () {
                closeWebsocket();
            }
        }
    }
</script>
</body>
</html>
```

上述代码中，在 head 标签中导入项目所需要的 js 库，如 jquery.js 和 echarts.js，设置页面编码格式等；在 div 标签中定义要绘制的图表的样式，包括 5 个图表；在 script 标签中 echartShowa()方法实现图表数据的渲染，在实现数据渲染时要确保 id 与 div 标签中定义的 id 一致，例如渲染 2023 年月均日照时长数据时，echartShowa()方法中 ECharts 初始化的元素 id 为 sunshineDiv，需要与各 div 中定义的 id 一致，否则数据渲染将失败。

3. 配置 Tomcat

参照 5.1.2 节第(6)步配置项目 Tomcat，配置完成后即可启动项目，项目启动成功部分截图如图 7-3 所示。

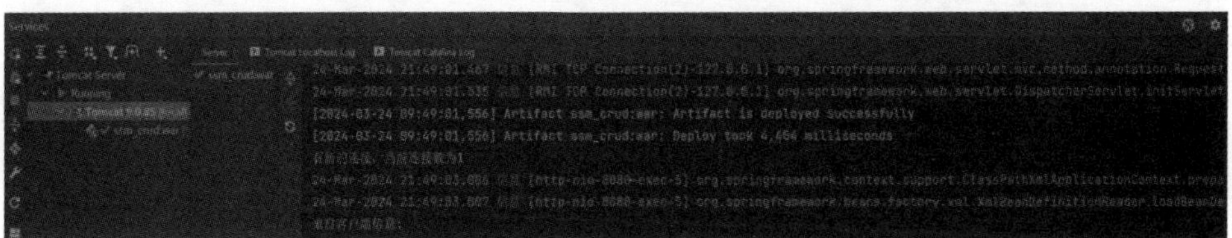

图 7-3　Tomcat 启动截图

7.6　运 行 结 果

项目运行成功，从不同维度绘制出可视化图表，图 7-4 是日照时长分析图，图 7-5 是温度分析图，图 7-6 是风速分析图，图 7-7 是降雨量分析图，图 7-8 是小麦产量分析图。

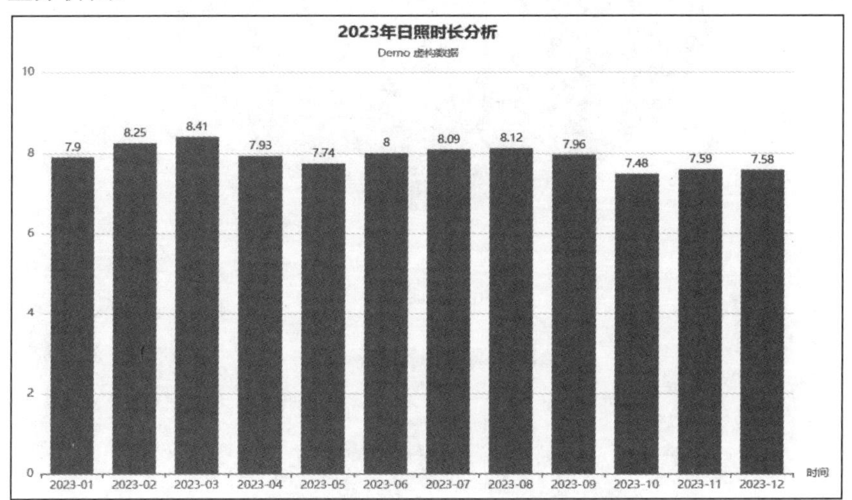

图 7-4　日照时长分析图

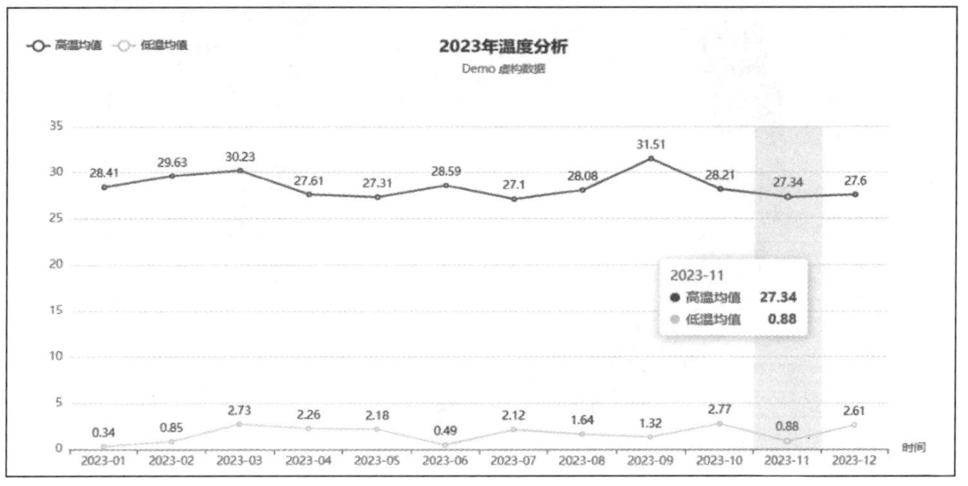

图 7-5　温度分析图

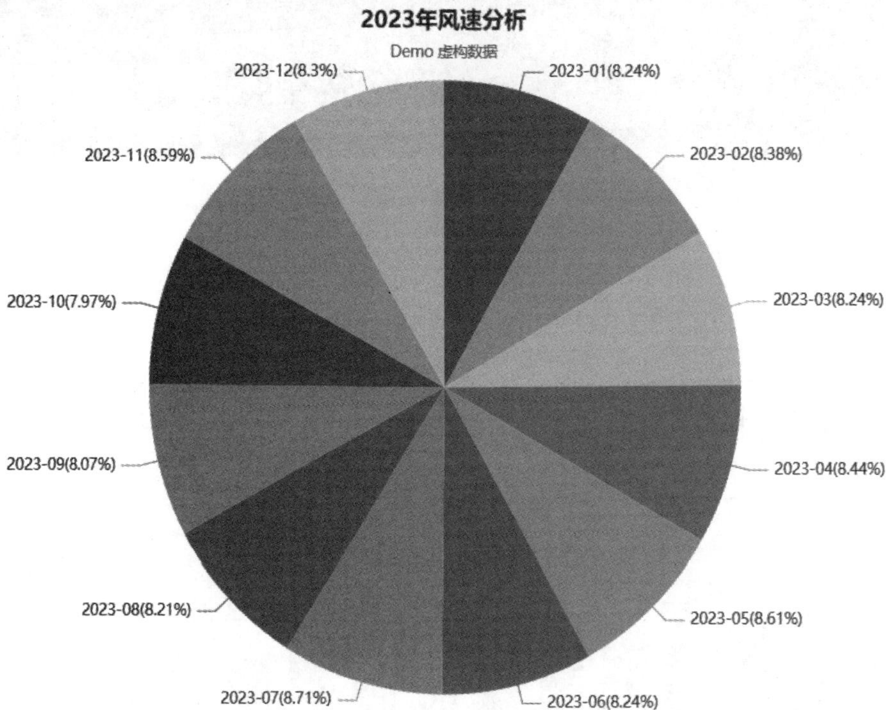

图 7-6 风速分析图

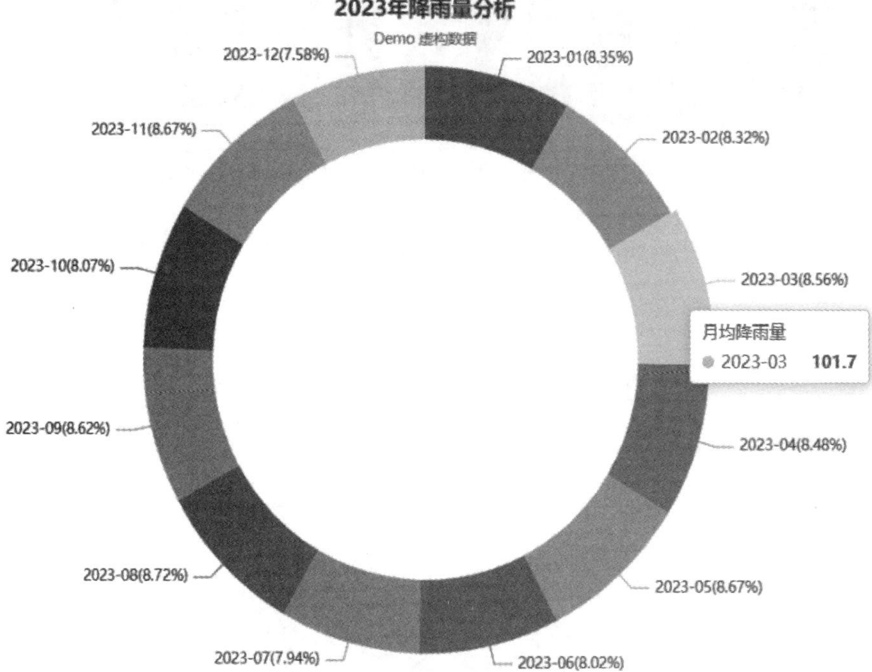

图 7-7 降雨量分析图

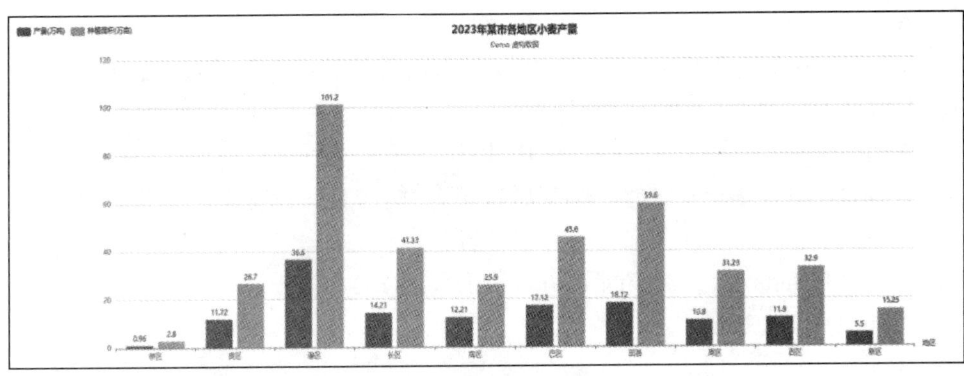

图 7-8　小麦产量分析图

企业通过这些图表能够直观地了解各地区的农业生产状况，因此完全实现了项目功能需求。

强化练习

1. 使用 MyBatis 持久化框架进行数据查询需要返回一个实体类的集合，在<select>标签中需要定义的一个属性是(　　)。

A. List

B. resultMap

C. HashMap

D. Tree

2. WebSocket 与 HTTP 的主要区别是(　　)。

A. WebSocket 基于 TCP 协议，HTTP 基于 UDP 协议

B. WebSocket 支持持久连接，HTTP 不支持

C. WebSocket 不支持发送二进制数据，HTTP 支持

D. WebSocket 不支持加密传输，HTTP 支持

3. 在 ECharts 的 legend 配置项中，itemGap 的默认间隔值是(　　)。

A. 0　　B. 5

C. 10　　D. 20

4. 在 WebSocket 中，(　　)事件会在接收到服务器消息时触发。

A. onopen

B. onmessage

C. onclose

D. onerror

5. ECharts 仅支持在浏览器环境中运行。(　　)

A. 正确　　　　B. 错误

6. WebSocket 是一种无状态的协议，每次通信都需要携带完整的上下文信息。(　　)

A. 正确　　　　B. 错误

7. WebSocket 是一种安全的通信协议，可以确保数据的机密性和完整性。(　　)

A. 正确　　　　B. 错误

8. 在 MyBatis 中，(　　)配置可以用于数据库连接。

A. driverClassName

B. url

C. username

D. password

9. ECharts 支持的可视化类型有(　　)。

A. 折线图

B. 柱状图

C. 饼图

D. 地图

10. Spring 框架中依赖注入(DI)的作用是(　　)。

A. 提高代码的运行效率

B. 减小代码的体积

C. 实现控制反转，降低组件之间的耦合度

D. 提供自动化的数据库连接管理

进一步学习建议

完成智慧农业数据分析平台开发后，为了进一步拓宽知识面，提升专业技能，可以进一步学习以下内容：

(1) Websocket 的拓展应用：学习 Websocket 在实时数据更新、音视频聊天、在线游戏等场景中的应用。

(2) Vue.js 框架：该框架是一款用于构建用户界面的 JavaScript 框架。它基于标准 HTML、CSS 和 JavaScript 构建，并提供了一套声明式的、组件化的编程模型，高效地开发用户界面。

(3) SpringBoot 框架：该框架是基于 Spring 4.0 设计的，是对 SSM 框架的进一步封装和简化。使用 SpringBoot 框架进行应用开发，能够使用 Spring 框架的所有优秀特性，同时还能够减少各种复杂的配置过程，降低各依赖包之间的冲突，增强系统的稳定性。

考核评价

<table>
<tr><td colspan="5">考核评价表</td></tr>
<tr><td colspan="2">姓名</td><td></td><td>班　级</td><td></td></tr>
<tr><td colspan="2">学号</td><td></td><td>考评时间</td><td></td></tr>
<tr><td colspan="2">评价主题及总分</td><td colspan="2">评价内容及分数</td><td>评分</td></tr>
<tr><td rowspan="3">1</td><td rowspan="3">知识考核
(30)</td><td colspan="2">掌握 ECharts 常用组件和图表(柱状图、折线图、饼图、散点图、气泡图)的功能和使用方法(10 分)</td><td></td></tr>
<tr><td colspan="2">掌握 SSM 框架技术的使用方法，正确完成各种核心文件的创建与配置(10 分)</td><td></td></tr>
<tr><td colspan="2">掌握 WebSocket 通信协议，完成前端页面与后端接口数据的交互功能(10 分)</td><td></td></tr>
<tr><td rowspan="4">2</td><td rowspan="4">技能考核
(40)</td><td colspan="2">具备业务需求分析、功能设计、编码及测试的综合能力(10 分)</td><td></td></tr>
<tr><td colspan="2">使用 ECharts 完成不同场景的数据可视化(10 分)</td><td></td></tr>
<tr><td colspan="2">使用 SSM 框架及时完成业务模块接口服务(10 分)</td><td></td></tr>
<tr><td colspan="2">实现业务需求中各功能模块的开发，保质保量完成任务(10 分)</td><td></td></tr>
<tr><td rowspan="3">3</td><td rowspan="3">思政考核
(30)</td><td colspan="2">能独立解决问题，分析问题，帮助别人解决难题，在团队合作中乐于分享知识与经验(10 分)</td><td rowspan="3"></td></tr>
<tr><td colspan="2">具备团队合神和主动合作意识，服从分配与管理，有较强的责任心和良好的沟通能力(10 分)</td></tr>
<tr><td colspan="2">能编写高可读性、高质量代码，代码符合规范，注释清晰明了(10 分)</td></tr>
<tr><td colspan="4">评语：</td><td>汇总：</td></tr>
</table>

参 考 文 献

[1] 黑马程序员. MySQL 数据库入门[M]. 2 版. 北京：清华大学出版社，2022.

[2] 陈巧莉. 软件工程项目实践教程[M]. 4 版. 大连：大连理工大学出版社，2023.

[3] 党翠萍，谢小光，黄以宝. Web 前端开发案例与实战[M]. 上海：上海交通大学出版社，2022.

[4] 黑马程序员. Java Web 程序设计任务教程[M]. 2 版. 北京：人民邮电出版社，2021.

[5] 黑马程序员. Java EE 企业级应用开发教程[M]. 2 版. 北京：人民邮电出版社，2021.

[6] 郭立文，王洪波. Web 前端开发项目化教程[M]. 北京：中国水利水电出版社，2021.

[7] 黑马程序员. Java 基础案例教程[M]. 2 版. 北京：人民邮电出版社，2021.

[8] 黑马程序员. Spark 大数据分析与实战[M]. 北京：清华大学出版社，2019.

[9] 石坤泉，汤双霞. MySQL 数据库任务驱动式教程[M]. 3 版. 北京：人民邮电出版社，2022.

[10] 程显毅. 大数据技术导论[M]. 北京：机械工业出版社，2019.

[11] 范路桥，张良均. Web 数据可视化[M]. 北京：人民邮电出版社，2021.

[12] 克雷格 • 沃斯. Spring 实战[M]. 5 版. 张卫滨，译. 北京：人民邮电出版社，2020.

[13] 程乐，郑丽萍，刘万辉. JavaScript 程序设计实例教程[M]. 2 版. 北京：机械工业出版社，2020.

[14] 顾理琴，常村红，刘万辉. HTML5 + CSS3 网页设计与制作基础教程[M]. 北京：机械工业出版社，2018.

[15] 郭立文，郑赢，王海龙. 大数据应用开发实战[M]. 北京：中国铁道出版社，2023.